BIOLOGY: LIFE ON EARTH

BIOLOGY: LIFE ON EARTH

C.S. Chandoliya

CYBER TECH PUBLICATIONS
4264/3, Ansari Road, Daryaganj, New Delhi-110002 (India)
Ph.: 011-23244078. Fax: 011-23280028
E-Mail: cyberpublicationsdelhi@yahoo.com

BIOLOGY : LIFE ON EARTH

C.S. Chandoliya

First Edition 2009

Published by :

G.S. Rawat for **Cyber Tech Publications**
4264/3, Ansari Road, Daryaganj, New Delhi-110002 (India)
Ph.: 011-23244078. Fax: 011-23280028
E-Mail: cyberpublicationsdelhi@yahoo.com

Printed at : **Amit Enterprises**, Maujpur, Delhi, (9810678184)

Preface

The tremendous success of science in explaining natural phenomena and fostering technological innovation arises from its focus on explanations that can be inferred from confirmable data. Scientists seek to relate one natural phenomenon to another and to recognize the causes and effects of phenomena. In this way, they have developed explanations for the changing of the seasons, the movements of the sun and stars, the structure of matter, the shaping of mountains and valleys, the changes in the positions of continents over time, the history of life on Earth, and many other natural occurrences. By the same means, scientists have also deciphered which substances in our environment are harmful to humans and which are not, developed cures for diseases, and generated the knowledge needed to produce innumerable labor-saving devices.

The concept of biological evolution is one of the most important ideas ever generated by the application of scientific methods to the natural world. The evolution of all the organisms that live on Earth today from ancestors that lived in the past is at the core of genetics, biochemistry, neurobiology, physiology, ecology, and other biological disciplines. It helps to explain the emergence of new infectious diseases, the development of antibiotic resistance in bacteria, the agricultural relationships among wild and domestic plants and animals, the composition of Earth's atmosphere, the molecular machinery of the cell, the similarities between human beings and other primates, and countless other features of the biological and physical world.

— *Author*

CONTENTS

1

INTRODUCTION

Biology (from Greek *Bio* meaning life and *logos* meaning the study of) is the study of life. Biology encompasses a broad spectrum of academic fields that are often viewed as independent disciplines. However, together they address phenomena related to living organisms (biological phenomena) over a wide range of scales, from biophysics to ecology. It is concerned with the characteristics, classification, and behaviors of organisms, how species come into existence, and the interactions they have with each other and with the natural environment. There are however unifying features. All biological phenomena are subject to the same laws that other branches of science obey, such as the laws of thermodynamics and conservation of mass. Practically the continued existence of any biological organism is always dependant on acquiring sufficient energy to overcome deleterious entropic effects.

Many of the subdisciplines of biology, such as botany, zoology, and medicine are ancient. However, biology as a unified science first developed in the nineteenth century, as scientists discovered that all living things shared certain fundamental characteristics and were best studied as a whole. Today biology is one of the most prominent scientific fields. Over a million papers are published annually in a wide array of biology and medicine journals , and biology is a standard subject of instruction at schools and universities around the world.

As such a vast field, biology is divided into a number of subdisciplines. The old division by type of organism remains with botany encompassing the study of plants, zoology the study of animals, microbiology the study of microorganisms, and other similar disciplines. The field is also divided by the scale being studied: molecular biology looks at the fundamental chemistry of life; cellular biology looks a the basic building block of all life the cell; Physiology looks at the internal structure of organism; and ecology looks at how various organisms interrelate.

PRINCIPLES

Biology is a branch of science employing the scientific method to characterize and investigate knowledge. Scientific theories are based on scientific observations, and these theories are sometimes refined as new scientific information is compiled. Scientific theories can also be used to predict phenomena that has not yet been observed. Biological systems are sometimes statistically modeled, but as in other branches of science, theories are not always described using mathematics. The biological sciences are characterized and unified by several major underlying principles and concepts: universality, evolution, diversity, continuity, genetics, homeostasis, and interactions.

UNIVERSALITY

While organisms may vary immensely in appearance, habitat, and behaviour it is a central principle of biology that all life shares certain universal fundamentals. All known life has a carbon-based biochemistry, carbon is the fundamental building block of the molecules that make up all living things. Similarly water is the basic solvent for all known living organisms. While all these things are true of all organisms observed on Earth, in theory alternative forms of life could exist and some scientists do look at alternative biochemistry. All terrestrial organism use DNA and RNA-based genetic mechanisms to hold genetic information. Another universal principle is that all observed organisms with the exception of viruses are made of cells. Similarly, all organisms share common developmental processes.

EVOLUTION

The central organizing concept in biology is that all life has a common origin and has changed and developed through the process of evolution. This has led to the striking similarity of units and processes discussed in the previous section. Charles Darwin established evolution as a viable theory by articulating its driving force, natural selection (Alfred Russel Wallace is recognized as the co-discoverer of this concept). Genetic drift was embraced as an additional mechanism of evolutionary development in the modern synthesis of the theory.

The evolutionary history of a species-which describes the characteristics of the various species from which it descended-together with its genealogical relationship to every other species is called its phylogeny. Widely varied approaches to biology generate information about phylogeny. These include the comparisons of DNA sequences conducted within molecular biology or genomics, and comparisons of fossils or other records of ancient organisms in paleontology. Biologists organize and analyze evolutionary relationships through various methods, including phylogenetics, phenetics, and cladistics (The major events in the evolution of life, as biologists currently understand them, are summarized on this evolutionary timeline).

Ever since its articulation by Darwin and Wallace, the theory of evolution by natural selection has come under attack by people who disagree with scientific findings or interpretations regarding the origins and diversity of life, generally favoring instead religious explanations.

Early Mesopotamian, Chinese and Indian Biology

From early times, perhaps predating the appearance of modern humans, people must have had and passed on knowledge about plants and animals to increase their chances of survival. For example, they had to know how to avoid (or sometimes use) poisonous plants and animals and how to track, capture, and butcher different species of animals. They had to know which plants could be prepared to make good nets or baskets. In this sense, biology predates the written history of humans.

Agriculture requires specialised knowledge on plants and animals. Ancient Oriental people knew about the pollination of date palm from a very early point of time. In Mesopotamia they knew that pollen could be used in fertilizing plants. A business contract of the Hammurabi period (c. 1800 BCE) mentions flowers of the date palm as an article of commerce. In India texts described some aspects of bird life. In Egypt the metamorphosis of insects and frogs was described. Egyptians and Babylonians also knew of anatomy and physiology in various forms. In Mesopotamia, animals were sometimes kept in what can be described as the first zoological gardens.

However, superstition often blended with facts. In Babylon and Assyria organs of animals were used in prediction, and in Egypt medicine included a large amount of mysticism.

Classical Greek Biology

In the Ancient Greek and Hellenistic worlds scholars became more interested in empiricism. Aristotle is one of the most prolific natural philosophers of Antiquity. Though his early scientific work was speculative, Aristotle's later biological works demonstrate great concern for observation. Aristotle did not experiment, however, holding that items display their real natures in their own environments, rather than controlled artificial ones. While in physics and chemistry, this assumption has been found unhelpful, in zoology and ethology it has not, and Aristotle's work "retains real interest".

He made countless observations of nature, especially the habits and attributes of plants and animals in the world around him, which he devoted considerable attention to categorizing. In all, Aristotle classified 540 animal species, and dissected at least 50. Aristotle believed that intellectual purposes, formal causes, guided all natural processes. Such a teleological view gave Aristotle cause to justify observed data as an expression of formal design. Noting that "no animal has, at the same time, both tusks and horns", and "a single-hooved animal with two horns I have never seen", Aristotle suggested that Nature, giving no animal both horns and tusks, was staving off vanity, and giving creatures faculties only to such a degree as they are necessary. Noting that ruminants had a multiple stomach and weak teeth, he supposed the first was to compensate for the latter, with Nature trying to preserve type of balance.

In a similar fashion, Aristotle believed that creatures were arranged in a graded scale of perfection rising from plants on up to man. His system had eleven grades, arranged according "to the degree to which they are infected with potentiality", expressed in their form at birth. The highest animals laid warm and wet creatures alive, the lowest bore theirs cold, dry, and in thick eggs. Aristotle also noted that the level of a creature's perfection was reflected in its form, but not foreordained by that form. Also important, in the view of Aristotle, was the quality of that creature's soul. Dividing souls into three groups, Aristotle noted that plants possessed a vegetative soul, responsible for reproduction and growth; animals a vegetative and a sensitive soul, responsible for mobility and sensation; and humans a vegetative, a sensitive, and a rational soul, capable of thought and reflection. Aristotle, in contrast to earlier philosophers, placed the rational soul in the heart, rather than the brain. Notable is Aristotle's division of sensation and thought, which generally went against previous philosophers, with the exception of Alcmaeon of Croton.

Aristotle's successor at the Lyceum, Theophrastus, wrote a series of books on botany, the History of Plants, which survived as the most important contribution of antiquity to botany, even into the Middle Ages. Many of Theophrastus' names survive into modern times, such as carpos for fruit, and pericarpion for seed vessel. Rather than focus on formal causes, as Aristotle did, Theophrastus suggested a mechanistic scheme, drawing analogies between natural and artificial processes, and relying on Aristotle's concept of the efficient cause. Theophrastus also recognized the role of sex in the reproduction of some higher plants, though this last discovery was lost in later ages.

Hellenistic Biology

Following Theophrastus, the Lyceum failed to produce any original work. Though interest in Aristotle's ideas survived, they were generally taken unquestioningly. It is not until the age of Alexandria under the Ptolemies that advances in biology can be again found. The first medical teacher at Alexandria was Herophilus of Chalcedon, who corrected Aristotle, placing intelligence in the brain, and connected the nervous system to motion and sensation. Herophilus also distinguished between veins and arteries, noting that the latter pulse while the former do not. In the same vein, he developed a diagnostic technique which relied upon distinguishing different types of pulse. He, and his contemporary, Erasistratus of Chios, researched the role of veins and nerves, mapping their courses across the body.

Erasistratus connected the increased complexity of the surface of the human brain compared to other animals to its superior intelligence. He sometimes employed experiments to further his research, at one time repeatedly weighing a caged bird, and noting its weight loss between feeding times. Following his teacher's researches into pneumatics, he claimed that the human system of blood vessels was controlled by vacuums, drawing blood across the body. In Erisistratus' physiology, air enters the body, is then drawn by the lungs into the heart, where it is transformed into vital spirit, and is then pumped by the arteries throughout the body. Some of this vital spirit reaches the brain, where it is transformed into animal spirit, which is then distributed by the nerves. Herophilus and Erasistratus performed their experiments upon criminals given them by their Ptolemaic kings. They dissected these criminals alive, and "while they were still breathing they observed parts which nature had formerly concealed, and examined their position, colour, shape, size, arrangement, hardness, softness, smoothness, connection." In ancient Rome, Pliny the Elder was known for his knowledge of plants and nature. Later, Claudius Galen became a pioneer in medicine and anatomy.

Medieval Knowledge

The decline of the Roman Empire led to the disappearance or destruction of much knowledge. This time is often called the dark ages. However, some people who dealt with medical issues still studied plants and animals as well. In Byzantium and the Islamic world, natural philosophy was kept alive. Many of the Greek works were translated into Arabic and many of the works of Aristotle were preserved. Of the Arab biologists, al-Jahiz, who died about 868, is particularly noteworthy. He wrote Kitab al Hayawan (Book of animals). In the 1200s the German scholar named Albertus Magnus wrote De vegetabilibus, seven books, and De animalibus, 26 books. He was particularly interested in plant propagation and reproduction and discussed in some detail the sexuality of plants and animals. He was also one the teachers of Thomas Aquinas.

Persian and Islamic Biology

Persia and other Islamic areas became important in the development of science. Based on Greek and Indian science and connected to Europe they were in a good position to help science develop. There were also Arab and Turkish scientists but the most important ones were Persians. Avicenna (commemorated in the genus Avicennia) was very important in biology and recorded many findings. He is sometimes regarded as the father of modern medicine. Rhazes was also very important and he was also a good biologist.

The Renaissance

Interestingly, as many visual artists were interested in the bodies of animals and humans, they studied the physiology in detail. Such comparisons as that between a horse leg and a human leg were made. Otto Brunfels, Hieronymus Bock (also known as Hieronymus Tragus) and Leonhart Fuchs were three men who wrote books about wild plants; they have been referred to as the fathers of German botany. Books about animals were also made, such as those by Conrad Gesner, illustrated by, among others, Albrecht Dürer.

Seventeenth and Eighteenth Century Developments

In 1628 William Harvey explained that blood circulates throughout the body, and is pumped by the heart. Antony van Leeuwenhoek's invention of the microscope in about 1650 opened up the micro-world of biology. The History of Plants was greatly extended, almost into an encyclopedia, by Giovanni Bodeo da

Stapel in 1644 AD. In 1658 Jan Swammerdam was the first to observe and describe red blood cells, while Leeuwenhoek was the first to describe spermatozoa, bacteria and infusoria in the 1670s and 1680s. By the 1690s plants were, like animals, known to be sexual, having stamens and pistils.

In 1669 Nicholas Steno published an essay on how the remains of living organisms could be trapped in layers of sediment and mineralized to produce fossils. Although Steno's ideas about fossilization were well known and much debated among natural philosophers, an organic origin for all fossils would not be accepted by all naturalists until the end of the 18th century due to philosophical and theological debate about issues such as the age of the earth, the role of the Noachian flood, and extinction.

Systematizing, naming and classifying dominated biology throughout much of the 17th and 18th centuries. Carolus Linnaeus published a basic taxonomy for the natural world in 1735, and in the 1750s, introduced scientific names for all his species. The discovery and description of new species, and collecting specimens became a widespread passion of scientific gentlemen.

Nineteenth Century: the Emergence of Biological Disciplines

Up through the nineteenth century, the scope of biology was largely divided between medicine, which investigated questions of form and function (i.e., physiology), and natural history, which was concerned with the diversity of life and interactions among different forms of life and between life and non-life. By 1900, much of these domains overlapped, while natural history and (and its counterpart natural philosophy) had largely given way to more specialized scientific disciplines-cytology, bacteriology, morphology, embryology, geography, and geology.

DIVERSITY

Classification is the province of the disciplines of systematics and taxonomy. Taxonomy places organisms in groups called taxa, while systematics seeks to define their relationships with each other. This classification technique has evolved to reflect advances in cladistics and genetics, shifting the focus from physical similarities and shared characteristics to phylogenetics. Traditionally, living things have been divided into five kingdoms:

Monera--Protista--Fungi--Plantae--Animalia

However, many scientists now consider this five-kingdom system to be outdated. Modern alternative classification systems generally begin with the three-domain system:

Archaea (originally Archaebacteria)--Bacteria (originally Eubacteria)--Eukaryota

These domains reflect whether the cells have nuclei or not, as well as differences in the cell exteriors.

Further, each kingdom is broken down continuously until each species is separately classified. The order is 1) Kingdom, 2) Phylum, 3) Class, 4) Order, 5) Family, 6) Genus, 7) Species. The scientific name of an organism is obtained from its Genus and Species. For example, humans would be listed as Homo sapiens. Homo would be the Genus and Sapiens is the species. Whenever writing the scientific name of an organism it is proper to capitalize the first letter in the genus and put all of the species in lowercase; in addition the entire term would be put in italics. The term used for classification is called Taxonomy.

There is also a series of intracellular parasites that are progressively "less alive" in terms of metabolic activity:

Viruses--Viroids--Prions

CONTINUITY

Up into the 19th century, it was commonly believed that life forms could appear spontaneously under certain conditions. This misconception was challenged by William Harvey's diction that "all life [is] from [an] egg" (from the Latin "Omne vivum ex ovo"), a foundational concept of modern biology. It simply means that there is an unbroken continuity of life from its initial origin to the present time.

A group of organisms shares a common descent if they share a common ancestor. All organisms on the Earth have been and are descended from a common ancestor or an ancestral gene pool. This last universal common ancestor of all organisms is believed to have appeared about 3.5 billion years ago. Biologists generally regard the universality of the genetic code as definitive evidence in favor of the theory of universal common descent (UCD) for all bacteria, archaea, and eukaryotes.

HOMEOSTASIS

Homeostasis is the ability of an open system to regulate its internal environment to maintain a stable condition by means of multiple dynamic equilibrium adjustments controlled by interrelated regulation mechanisms. All living organisms, whether unicellular or multicellular, exhibit homeostasis. Homeostasis manifests itself at the cellular level through the maintenance of a stable internal acidity (pH); at the organismic level, warm-blooded animals maintain a constant internal body temperature; and at the level of the ecosystem, as when atmospheric carbon dioxide levels rise and plants are theoretically able to grow healthier and remove more of the gas from the atmosphere. Tissues and organs can also maintain homeostasis.

Interactions

Every living thing interacts with other organisms and its environment. One reason that biological systems can be difficult to study is that so many different interactions with other organisms and the environment are possible, even on the smallest of scales. A microscopic bacterium responding to a local sugar gradient is responding to its environment as much as a lion is responding to its environment when it searches for food in the African savannah. For any given species, behaviors can be co-operative, aggressive, parasitic or symbiotic. Matters become more complex when two or more different species interact in an ecosystem. Studies of this type are the province of ecology.

Scope

Biology has become such a vast research enterprise that it is not generally regarded as a single discipline, but as a number of clok but i doich assist in understanding the genetic variation of a population; and physiology borrows extensively from cell biology in describing the function of organ systems. Ethology and comparative psychology extend biology to the analysis of animal behavior and mental characteristics, whilst Evolutionary psychology proposes that the field of psychology, including in regard to humans, is a branch of biology.

STRUCTURE OF LIFE

Molecular biology is the study of biology at a molecular level. This field overlaps with other areas of biology, particularly with genetics and biochemistry. Molecular biology chiefly concerns itself with understanding the interactions between the various systems of a cell, including the interrelationship of DNA, RNA, and protein synthesis and learning how these interactions are regulated.

Cell biology studies the physiological properties of cells, as well as their behaviors, interactions, and environment. This is done both on a microscopic and molecular level. Cell biology researches both single-celled organisms like bacteria and specialized cells in multicellular organisms like humans.

Understanding cell composition and how they function is fundamental to all of the biological sciences. Appreciating the similarities and differences between cell types is particularly important in the fields of cell and molecular biology. These fundamental similarities and differences provide a unifying theme, allowing the principles learned from studying one cell type to be extrapolated and generalized to other cell types.

Genetics is the science of genes, heredity, and the variation of organisms. In modern research, genetics provides important tools in the investigation of the function of a particular gene, or the analysis of genetic interactions. Within organisms, genetic information generally is carried in chromosomes, where it is represented in the chemical structure of particular DNA molecules. Genes encode the information necessary for synthesizing proteins, which in turn play a large role in influencing (though, in many instances, not completely determining) the final phenotype of the organism.

Developmental biology studies the process by which organisms grow and develop. Originating in embryology, modern developmental biology studies the genetic control of cell growth, differentiation, and "morphogenesis," which is the process that gives rise to tissues, organs, and anatomy. Model organisms for developmental biology include the round worm Caenorhabditis elegans, the fruit fly Drosophila melanogaster, the zebrafish Brachydanio rerio, the mouse Mus musculus, and the weed Arabidopsis thaliana.

PHYSIOLOGY OF ORGANISMS

Physiology studies the mechanical, physical, and biochemical processes of living organisms by attempting to understand how all of the structures function as a whole. The theme of "structure to function" is central to biology. Physiological studies have traditionally been divided into plant physiology and animal physiology, but the principles of physiology are universal, no matter what particular organism is being studied. For example, what is learned about the physiology of yeast cells can also apply to human cells. The field of animal physiology extends the tools and methods of human physiology to non-human species. Plant physiology also borrows techniques from both fields.

Anatomy is an important branch of physiology and considers how organ systems in animals, such as the nervous, immune, endocrine, respiratory, and circulatory systems, function and interact. The study of these systems is shared with medically oriented disciplines such as neurology and immunology.

DIVERSITY AND EVOLUTION OF ORGANISMS

Evolutionary biology is concerned with the origin and descent of species, as well as their change over time, and includes scientists from many taxonomically-oriented disciplines. For example, it generally involves scientists who have special training in particular organisms such as mammalogy, ornithology, or herpetology, but use those organisms as systems to answer general questions about evolution. Evolutionary biology is mainly based on paleontology, which uses the fossil record to answer questions about the mode and tempo of evolution, as well as the developments in areas such as population genetics and evolutionary theory. In the 1990s, developmental biology re-entered evolutionary biology from its initial exclusion from the modern synthesis through the study of evolutionary developmental biology. Related fields which are often considered part of evolutionary biology are phylogenetics, systematics, and taxonomy.

The two major traditional taxonomically-oriented disciplines are botany and zoology. Botany is the scientific study of plants. Botany covers a wide range of scientific disciplines that study the growth, reproduction, metabolism, development, diseases, and evolution of plant life. Zoology involves the study of animals, including the study of their physiology within the fields of anatomy and embryology. The common genetic and developmental mechanisms of animals and plants is studied in molecular biology, molecular genetics, and developmental biology. The ecology of animals is covered under behavioral ecology and other fields.

CLASSIFICATION OF LIFE

The dominant classification system is called Linnaean taxonomy, which includes ranks and binomial nomenclature. How organisms are named is governed by international agreements such as the International Code of Botanical Nomenclature (ICBN), the International Code of Zoological Nomenclature (ICZN), and the International Code of Nomenclature of Bacteria (ICNB). A fourth Draft BioCode was published in 1997 in an attempt to standardize naming in these three areas, but it has yet to be formally adopted. The Virus cInternational Code of Virus Classification and Nomenclature (ICVCN) remains outside the BioCode.

INTERACTIONS OF ORGANISMS

Ecology studies the distribution and abundance of living organisms, and the interactions between organisms and their environment. The environment of an organism includes both its habitat, which can be described as the sum of local abiotic factors such as climate and geology, as well as the other the organisms that share its habitat. Ecological systems are studied at several different levels, from individuals and populations to ecosystems and the biosphere. As can be surmised, ecology is a science that draws on several disciplines. Ethology studies animal behavior (particularly of social animals such as primates and canids), and is sometimes considered a branch of zoology. Ethologists have been particularly concerned with the evolution of behavior and the understanding of behavior in terms of the theory of natural selection. In one sense, the first modern ethologist was Charles Darwin, whose book The expression of the emotions in animals and men influenced many ethologists.

Biogeography studies the spatial distribution of organisms on the Earth, focusing on topics like plate tectonics, climate change, dispersal and migration, and cladistics.

ETYMOLOGY

Formed by combining the Greek *bios*, meaning 'life', and *logos*, meaning 'study of', the word "biology" in its modern sense seems to have been introduced independently by Gottfried Reinhold Treviranus (Biologie oder Philosophie der lebenden Natur, 1802) and by Jean-Baptiste Lamarck (Hydrogéologie, 1802). The word itself is sometimes said to have been coined in 1800 by Karl Friedrich Burdach, but it appears in the title of Volume 3 of Michael Christoph Hanov's Philosophiae naturalis sive physicae dogmaticae: Geologia, biologia, phytologia generalis et dendrologia, published in 1766.

History

Though the concept of biology as a single coherent field of knowledge only arose in the 19th century, the biological sciences emerged from traditions of medicine and natural history reaching back to the ancient Greeks (particularly Galen and Aristotle, respectively). During the Renaissance and Age of Discovery, renewed interest in empiricism as well as the rapidly increasing number of known organisms led to significant developments in biological thought; Vesalius inaugurated the rise of experimentation and careful observation in physiology, and a series of naturalists culminating with Linnaeus and Buffon began to create a conceptual framework for analyzing the diversity of life and the fossil record, as well as the development and behavior of plants and animals. The growing importance of natural theology-partly a response to the rise of mechanical philosophy-was also an important impetus for the growth of natural history (though it also further entrenched the argument from design).

In the 18th century many fields of science-including botany, zoology, and geology-began to professionalize, forming the precursors of scientific disciplines in the modern sense (though the process would not be complete until the late 1800s). Lavoisier and other physical scientists began to connect the animate and inanimate worlds through the techniques and theory of physics and chemistry. Into the 19th century, explorer-naturalists such as Alexander von Humboldt tried to elucidate the interactions between organisms and their environment, and the ways these relationships depend on geography-creating the foundations for biogeography, ecology and ethology. Many naturalists began to reject essentialism and seriously consider the possibilities of extinction and the mutability of species. These developments, as well as the results of new fields such as embryology and paleontology, were synthesized in Darwin's theory of evolution by natural selection. The end of the 19th century saw debates over spontaneous generation and the rise of the germ theory of disease and the fields of cytology, bacteriology and physiological chemistry, though the problem of inheritance was still a mystery.

In the early 20th century, the rediscovery of Mendel's work led to the rapid development of genetics by Thomas Hunt Morgan and his students, and by the 1930s the combination of population genetics and natural selection led to the "neo-Darwinian synthesis" and the rise of the discipline of evolutionary biology. New biological disciplines developed rapidly, especially after Watson and Crick discovered the structure of DNA in 1953. Following the cracking of the genetic code and the establishment of the Central Dogma, biology has largely split between organismal biology-consisting of ecology, ethology, systematics, paleontology, evolutionary biology, developmental biology, and other disciplines that deal with whole organisms or groups of organisms-and the constellation of disciplines related to molecular biology-including cell biology, biophysics, biochemistry, neuroscience, immunology, and many other overlapping subjects. By the beginning of the 21st century, this schism had begun to heal. Organismal biologists increasingly turned to molecular techniques and styles of thought, and molecular and cell biologists increasingly focused on the interplay between genes and the natural environment, as well as the genetic heterogeneity of natural populations such as humans.

BIOLOGIST

A biologist is a scientist devoted to and producing results in biology through the study of organisms. Typically biologists study organisms and their relationship to their environment. Biologists involved in basic research attempt to discover underlying mechanisms that govern how organisms work. Biologists involved in applied research attempt to develop or improve medical, industrial or agricultural processes. Some biologists work in management positions coordinating or supervising research. Technicians perform specialized tasks using laboratory equipment. (U.S. Department of Labor.)

Types

There are many types of biologists. Some deal with the macroscopic world and other with the microscopic. There is much overlap between different fields, and it is often hard to classify a biologist as only one of them.

BIOCHEMISTRY

Biochemistry is the study of the chemical processes and transformations in living organisms. This article only discusses terrestrial biochemistry (carbon-and water-based), as all the life forms we know are on Earth. Since life forms alive today are hypothesized by some to have descended from the same common ancestor, they would naturally have similar biochemistries, even for matters that seem to be essentially arbitrary, such as handedness of various biomolecules. It is unknown whether alternative biochemistries are possible or practical.

Biochemistry is the study of the structure and function of cellular components, such as proteins, carbohydrates, lipids, nucleic acids, and other biomolecules. Chemical biology aims to answer many questions arising from biochemistry by using tools developed within synthetic chemistry.

Although there are a vast number of different biomolecules, many are complex and large molecules (called polymers) that are composed of similar repeating subunits (called monomers). Each class of polymeric biomolecule has a different set of subunit types. For example, a protein is a polymer made up of many amino acids. Biochemistry studies the chemical properties of important biological molecules, like proteins, in particular the chemistry of enzyme-catalyzed reactions (enzymes are a type of protein).

The biochemistry of cell metabolism and the endocrine system has been extensively described. Other areas of biochemistry include the genetic code (DNA, RNA), protein synthesis, cell membrane transport, and signal transduction.

History of Biochemistry

Originally, it was generally believed that life was not subject to the laws of science the way non-life was. It was thought that only living beings could produce the molecules of life (from other, previously existing biomolecules). Then, in 1828, Friedrich Wöhler published a paper about the synthesis of urea, proving that organic compounds can be created artificially. The dawn of biochemistry may have been the discovery of the first enzyme, diastase (today called amylase), in 1833 by Anselme Payen. Eduard Buchner contributed the first demonstration of a complex biochemical process outside of a cell in 1896: alcoholic fermentation in cell extracts of yeast. Although the term "biochemistry" seems to have been first used in 1881, it is generally accepted that the formal coinage of biochemistry occurred in 1903 by Carl Neuber, a German chemist. Since then, biochemistry has advanced, especially since the mid-20th century, with the development of new techniques such as chromatography, X-ray diffraction, NMR spectroscopy, radioisotopic labelling, electron microscopy and molecular dynamics simulations.

These techniques allowed for the discovery and detailed analysis of many molecules and metabolic pathways of the cell, such as glycolysis and the Krebs cycle (citric acid cycle). Today, the findings of biochemistry are used in many areas, from genetics to molecular biology and from agriculture to medicine.

Carbohydrates

The function of carbohydrates includes energy storage and providing structure. Sugars are carbohydrates, although there are carbohydrates that are not sugars. There are more carbohydrates on Earth than any other type of biomolecule.

Monosaccharides

The simplest type of carbohydrate is a monosaccharide, which among other properties contains carbon, hydrogen, and oxygen, mostly in a ratio of 1:2:1 (generalized formula CnH_2nOn, where n is at least 3). Glucose, one of the most important carbohydrates, is an example of a monosaccharide. So is fructose, the sugar that gives fruits their sweet taste. Some carbohydrates (especially after condensation to oligo-and polysaccharides) contain less carbon relative to H and O, which still are present in 2:1 (H:O) ratio. Monosaccharides can be grouped into aldoses (having an aldehyde group at the end of the chain, e.g. glucose) and ketoses (having a keto group in their chain; e.g. fructose).

Both aldoses and ketoses occur in an equilibrium between the open-chain forms and (starting with chain lengths of C_4) cyclic forms. These are generated by bond formation between one of the hydroxy groups of the sugar chain with the carbon of the aldehyde or keto group in a hemiacetal bond. This leads to saturated five-membered (in furanoses) or six-membered (in pyranoses) heterocyclic rings containing one O as heteroatom.

Disaccharides

Two monosaccharides can be joined together using dehydration synthesis, in which a hydrogen atom is removed from the end of one molecule and a hydroxyl group (-OH) is removed from the other; the remaining residues are then attached at the sites from which the atoms were removed. The H-OH or H_2O is then released as a molecule of water, hence the term dehydration.

The new molecule, consisting of two monosaccharides, is called a disaccharide and is conjoined together by a glycosidic or ether bond. The reverse reaction can also occur, using a molecule of water to split up a disaccharide and break the glycosidic bond; this is termed hydrolysis.

The most well-known disaccharide is sucrose, ordinary sugar (in scientific contexts, called table sugar or cane sugar to differentiate it from other sugars). Sucrose consists of a glucose molecule and a fructose molecule joined together. Another important disaccharide is lactose, consisting of a glucose molecule and a galactose molecule. As most humans age, the production of lactase, the enzyme that hydrolyzes lactose back into glucose and galactose, typically decreases. This results in lactase deficiency, also called lactose intolerance.

Sugar polymers are characterised by having reducing or non-reducing ends. A reducing end of a carbohydrate is a carbon atom which can be in equilibrium with the open-chain aldehyde or keto form. If the joining of monomers takes place at such a carbon atom, the free hydroxy group of the pyranose or furanose form is exchanged with an OH-side chain of another sugar, yielding a full acetal.

This prevents opening of the chain to the aldehyde or keto form and renders the modified residue non-reducing. Lactose contains a reducing end at its glucose moiety, whereas the galactose moiety form a full acetal with the C_4-OH group of glucose. Saccharose does not have a reducing end because of full acetal formation between the aldehyde carbon of glucose (C_1) and the keto carbon of fructose (C_2).

Oligosaccharides and Polysaccharides

When a few (around three to six) monosaccharides are joined together, it is called an oligosaccharide (oligo-meaning "few"). These molecules tend to be used as markers and signals, as well as having some other uses.

Many monosaccharides joined together make a polysaccharide. They can be joined together in one long linear chain, or they may be branched. Two of the most common polysaccharides are cellulose and glycogen, both consisting of repeating glucose monomers.

- Cellulose is made by plants and is an important structural component of their cell walls. Humans can neither manufacture nor digest it.
- Glycogen, on the other hand, is an animal carbohydrate; humans use it as a form of energy storage.

Use of Carbohydrates as an Energy Source

Glucose is the major energy source in most life forms. For instance, polysaccharides are broken down into their monomers (glycogen phosphorylase removes glucose residues from glycogen). Disaccharides like lactose or sucrose are cleaved into their two component monosaccharides.

GLYCOLYSIS (ANAEROBIC)

Glucose is mainly metabolized by a very important and ancient ten-step pathway called glycolysis, the net result of which is to break down one molecule of glucose into two molecules of pyruvate; this also produces a net two molecules of ATP, the energy currency of cells, along with two reducing equivalents in the form of converting NAD+ to NADH.

This does not require oxygen; if no oxygen is available (or the cell cannot use oxygen), the NAD is restored by converting the pyruvate to lactate (e.g. in humans) or to ethanol plus carbon dioxide (e.g. in yeast). Other monosaccharides like galactose and fructose can be converted into intermediates of the glycolytic pathway.

Aerobic

In aerobic cells with sufficient oxygen, like most human cells, the pyruvate is further metabolized. It is irreversibly converted to acetyl-CoA, giving off one carbon atom as the waste product carbon dioxide, generating another reducing equivalent as NADH. The two molecules acetyl-CoA (from one molecule of glucose) then enter the citric acid cycle, producing two more molecules of ATP, six more NADH molecules and two reduced (ubi)quinones (via FADH2 as enzyme-bound cofactor), and releasing the remaining carbon atoms as carbon dioxide. The produced NADH and quinol molecules then feed into the enzyme complexes of the respiratory chain, an electron transport system transferring the electrons ultimately to oxygen and conserving the released energy in the form of a proton gradient over a membrane (inner mitochondrial membrane in eukaryotes). Thereby, oxygen is reduced to water and the original electron acceptors NAD+ and quinone are regenerated. This is why humans breathe in oxygen and breathe out carbon dioxide. The energy released from transferring the electrons from high-energy states in NADH and quinol is conserved first as proton gradient and converted to ATP via ATP synthase. This generates an additional 28 molecules of ATP (24 from the 8 NADH + 4 from the 2 quinols), totaling to 32 molecules of ATP conserved per degraded glucose (two from glycolysis + two from the citrate cycle). It is clear that using oxygen to completely oxidize glucose provides an organism with far more energy than any oxygen-independent metabolic feature, and this is thought to be the reason why complex life appeared only after Earth's atmosphere accumulated large amounts of oxygen.

Gluconeogenesis

In vertebrates, vigorously contracting skeletal muscles (during weightlifting or sprinting, for example) do not receive enough oxygen to meet the energy demand, and so they shift to anaerobic metabolism, converting glucose to lactate (lactic acid). The liver regenerates the glucose, using a process called gluconeogenesis. This process is not quite the opposite of glycolysis, and actually requires three times the amount of energy gained from glycolysis (six molecules of ATP are used, compared to the two gained in glycolysis). Analogous to the above reactions, the glucose produced can then undergo glycolysis in tissues that need energy, be stored as glycogen (or starch in plants), or be converted to other monosaccharides or joined into di-or oligosaccharides.

Proteins

Like carbohydrates, some proteins perform largely structural roles. For instance, movements of the proteins actin and myosin ultimately are responsible for the contraction of skeletal muscle. One property many proteins have is that they specifically bind to a certain molecule or class of molecules-they may be extremely selective in what they bind. Antibodies are an example of proteins that attach to one specific type of molecule. In fact, the enzyme-linked immunosorbent assay (ELISA), which uses antibodies, is currently one of the most sensitive tests modern medicine uses to detect various biomolecules. Probably the most important proteins, however, are the enzymes. These amazing molecules recognize specific reactant molecules called substrates; they then catalyze the reaction between them. By lowering the activation energy, the enzyme speeds up that reaction by a rate of 1011 or more: a reaction that would normally take over 3,000 years to complete spontaneously might take less than a second with an enzyme. The enzyme itself is not used up in the process, and is free to catalyze the same reaction with a new set of substrates. Using various modifiers, the activity of the enzyme can be regulated, enabling control of the biochemistry of the cell as a whole.

In essence, proteins are chains of amino acids. An amino acid consists of a carbon atom bound to four groups. One is an amino group,-NH_2, and one is a carboxylic acid group,-COOH (although these exist as-NH^{3+} and-COO-under physiologic conditions). The third is a simple hydrogen atom. The fourth is commonly denoted "-R" and is different for each amino acid. There are twenty standard amino acids. Some of these have functions by themselves or in a modified form; for instance, glutamate functions as an important neurotransmitter. Amino acids can be joined together via a peptide bond. In this dehydration synthesis, a water molecule is removed and the peptide bond connects the nitrogen of one amino acid's amino group to the carbon of the other's carboxylic acid group. The resulting molecule is called a dipeptide, and short stretches of amino acids (usually, fewer than around thirty) are called peptides or polypeptides. Longer stretches merit the title proteins. As an example, the important blood serum protein albumin contains 585 amino acid residues.

The structure of proteins is traditionally described in a hierarchy of four levels. The primary structure of a protein simply consists of its linear sequence of amino acids; for instance, "alanine-glycine-tryptophan-serine-glutamate-asparagine-glycine-lysine-...". Secondary structure is concerned with local morphology. Some combinations of amino acids will tend to curl up in a coil called an a-helix; some of these can be seen in the hemoglobin schematic above. Tertiary structure is the entire three-dimensional shape of the protein. This shape is determined by the sequence of amino acids. In fact, a single change can change the entire structure.

The ß chain of hemoglobin contains 146 amino acid residues; substitution of the glutamate residue at position 6 with a valine residue changes the behavior of hemoglobin so much that it results in sickle-cell disease. Finally quaternary structure is concerned with the structure of a protein with multiple peptide subunits, like hemoglobin with its four subunits. Not all proteins have more than one subunit.

Ingested proteins are usually broken up into single amino acids or dipeptides in the small intestine, and then absorbed. They can then be joined together to make new proteins. Intermediate products of glycolysis, the citric acid cycle, and the pentose phosphate pathway can be used to make all twenty amino acids, and most bacteria and plants possess all the necessary enzymes to synthesize them. Humans and other mammals, however, can only synthesize half of them. They cannot synthesize isoleucine, leucine, lysine, methionine, phenylalanine, threonine, tryptophan, and valine. These are the essential amino acids, since it is essential to ingest them. Mammals do possess the enzymes to synthesize alanine, asparagine, aspartate, cysteine, glutamate, glutamine, glycine, proline, serine, and tyrosine, the nonessential amino acids. While they can synthesize arginine and histidine, they cannot produce it in sufficient amounts for young, growing animals, and so these are often considered essential amino acids.

If the amino group is removed from an amino acid, it leaves behind a carbon skeleton called an a-keto acid. Enzymes called transaminases can easily transfer the amino group from one amino acid (making it an a-keto acid) to another a-keto acid (making it an amino acid). This is important in the biosynthesis of amino acids, as for many of the pathways, intermediates from other biochemical pathways are converted to the a-keto acid skeleton, and then an amino group is added, often via transamination. The amino acids may then be linked together to make a protein.

A similar process is used to break down proteins. It is first hydrolyzed into its component amino acids. Free ammonia (NH3, existing as the ammonium ion NH4+) in blood) is toxic to life forms. A suitable method for excreting it must therefore exist. Different strategies have evolved in different animals, depending on the animals' needs. Unicellular organisms, of course, simply release the ammonia into the environment. Similarly, bony fish can release the ammonia into the water where it is quickly diluted. In general, mammals convert the ammonia into urea, via the urea cycle.

Lipids

The term lipid comprises a diverse range of molecules and to some extent is a catchall for relatively water-insoluble or nonpolar compounds of biological origin, including waxes, fatty acids, fatty-acid derived phospholipids, sphingolipids, glycolipids and terpenoids (eg. retinoids and steroids). Some lipids are linear aliphatic molecules, while others have ring structures. Some are aromatic, while others are not. Some are flexible, while others are rigid. Most lipids have some polar character in addition to being largely nonpolar. Generally, the bulk of their structure is nonpolar or hydrophobic ("water-fearing"), meaning that it does not interact well with polar solvents like water. Another part of their structure is polar or hydrophilic ("water-loving") and will tend to associate with polar solvents like water. This makes them amphiphilic molecules (having both hydrophobic and hydrophilic portions). In the case of cholesterol, the polar group is a mere-OH (hydroxyl or alcohol). In the case of phospholipids, the polar groups are considerably larger and more polar.

Nucleic Acids

A nucleic acid is a complex, high-molecular-weight biochemical macromolecule composed of nucleotide chains that convey genetic information. The most common nucleic acids are deoxyribonucleic acid (DNA) and ribonucleic acid (RNA). Nucleic acids are found in all living cells and viruses.

Nucleic acid, so called because of its prevalence in cellular nuclei, is the generic name of family of biopolymers. The monomers are called nucleotides, and each consists of three components: a nitrogenous heterocyclic base (either a purine or a pyrimidine), a pentose sugar, and a phosphate group. Different nucleic acid types differ in the specific sugar found in their chain (e.g. DNA or deoxyribonucleic acid contains 2-deoxyriboses). Also, the nitrogenous bases possible in the two nucleic acids are different: adenine, cytosine, and guanine are possible in both RNA and DNA, while thymine is possible only in DNA and uracil is possible only in RNA.

Relationship to Other "Molecular-scale" Biological Sciences

Researchers in biochemistry use specific techniques native to biochemistry, but increasingly combine these with techniques and ideas from genetics, molecular biology and biophysics. There has never been a hard-line between these disciplines in terms of content and technique, but members of each discipline have in the past been very territorial; today the terms molecular biology and biochemistry are nearly interchangeable. The following figure is a schematic that depicts one possible view of the relationship between the fields:

- Biochemistry is the study of the chemical substances and vital processes occurring in living organisms.
- Genetics is the study of the effect of genetic differences on organisms. Often this can be inferred by the absence of a normal component (e.g. one gene). The study of "mutants"- organisms which lack one or more functional components with respect to the so-called "wild type" or normal phenotype. Genetic interactions (epistasis) can often confound simple interpretations of such "knock-out" studies.
- Molecular biology is the study of molecular underpinnings of the process of replication, transcription and translation of the genetic material. The central dogma of molecular biology where genetic material is transcribed into RNA and then translated into protein, despite being an oversimplified picture of molecular biology, still provides a good starting point for understanding the field. This picture, however, is undergoing revision in light of emerging novel roles for RNA.
- Chemical Biology seeks to develop new tools based on small molecules that allow minimal perturbation of biological systems while providing detailed information about their function. Further, chemical biology employs biological systems to create non-natural hybrids between biomolecules and synthetic devices (for example emptied viral capsids that can deliver gene therapy or drug molecules).

CELL BIOLOGY

Cell biology (also called cellular biology or formerly cytology, from the Greek kytos, "container") is an academic discipline that studies cells. This includes their physiological properties, their structure, the organelles they contain, interactions with their environment, their life cycle, division and death. This is done both on a microscopic and molecular level. Cell biology research extends to both the great diversity of single-celled organisms like bacteria and the many specialized cells in multicellular organisms like humans. Knowing the composition of cells and how cells work is fundamental to all of the biological sciences. Appreciating the similarities and differences between cell types is particularly important to the fields of cell and molecular biology. These fundamental similarities and differences provide a unifying theme, allowing the principles learned from studying one cell type to be extrapolated and generalized to other cell types. Research in cell biology is closely related to genetics, biochemistry, molecular biology and developmental biology.

Processes

Movement of Proteins

Every cell typically contains hundreds of different kinds of macromolecules that function together to generate the behavior of the cell. Each type of protein is usually sent to a particular part of the cell. An important part of cell biology is investigation of molecular mechanisms by which proteins are moved to different places inside cells or secreted from cells.

Most proteins are synthesized by ribosomes in the cytoplasm. This process is also known as protein biosynthesis or simply protein translation. Some proteins, such as those to be incorporated in membranes (membrane proteins), are transported into the ER during synthesis and further processed in the Golgi apparatus. From the Golgi, membrane proteins can move to the plasma membrane, to other subcellular compartments or they can be secreted from the cell. The ER and Golgi can be thought of as the "membrane protein synthesis compartment" and the "membrane protein processing compartment", respectively. There is a semi-constant flux of proteins through these compartments. ER and Golgi-resident proteins associate with other proteins but remain in their respective compartments. Other proteins "flow" through the ER and Golgi to the plasma membrane. Motor proteins transport mebrane protein-containing vesicles along cytoskeletal tracks to distant parts of cells such as axon terminals.

Some proteins that are made in the cytoplasm contain structural features that target them for transport into mitochondria or the nucleus. Some mitochondrial proteins are made inside mitochondria and are coded for by mitochondrial DNA. In plants, chloroplasts also make some cell proteins.

Extracellular and cell surface proteins destined to be degraded can move back into intracellular compartments upon being incorporated into endocytosed vesicles. Some of these vesicles fuse with (lysosomes) where the proteins are broken down to their individual amino acids. The degradation of some membrane proteins begins while still at the cell surface when they are cleaved by secretases. Proteins that function in the cytoplasm are often degraded by proteasomes.

Other Cellular Processes

- Cell division-The origin of new cells.
- Cell signaling-Regulation of cell behavior by signals from outside.
- Active transport and Passive transport-Movement of molecules into and out of cells.
- Adhesion-Holding together cells and tissues.
- Transcription and mRNA splicing-gene expression.
- Cell movement: Chemotaxis, Contraction, cilia and flagella
- DNA repair and Cell death
- Metabolism: Glycolysis, respiration, Photosynthesis

Techniques

- Microscopy and Immunostaining
- Gene knockdown and Transfection
- Cell culture and Radioactive tracers
- PCR and In situ hybridization
- DNA microarray screens of gene expression

Purification of Cells and their Parts

Purification of cells and their parts is achieved in the following ways:

- Flow cytometry

- Cell fractionation
    - o Release of cellular organelles by disruption of cells.
    - o Separation of different organelles by centrifugation.
- Proteins extracted from cell membranes by detergents and salts or other kinds of chemicals.
- Immunoprecipitation

Some Structures Inside Cells

- Organelle-term used for major subcellular structures
- Chloroplast-key organelle for photosynthesis
- Cilia-motile microtubule-containing structures of eukaryotes
- Cytoplasm-contents of the main fluid-filled space inside cells
- Cytoskeleton-protein filaments inside cells
- Ribosome-RNA and protein complex required for protein synthesis in cells
- Endoplasmic reticulum-major site of membrane protein synthesis
- Flagella-motile structures of bacteria, archaea and eukaryotes
- Golgi apparatus-site of protein glycosylation in the endomembrane system
- Membrane lipid and protein barrier
- Lipid bilayer-fundamental organizational structure of cell membranes
- Vesicle-small membrane-bounded spheres inside cells
- Mitochondrion-major energy-producing organelle
- Nucleus-holds most of the DNA of eukaryotic cells

CYTOPATHOLOGY

Cytopathology is a branch of pathology that studies and diagnoses diseases on the cellular level. The most common use of cytopathology is the Pap smear, used to detect cervical cancer at an early treatable stage. Two methods of collecting cells for analysis are:

1. *Exfoliative Cytology*: Cells are extracted from fluid shed into the body cavities. For example, in pleural fluid, ascitic fluid, or in the case of the Pap smear, cells scraped from the cervix.
2. *Fine Needle Aspiration Cytology or Needle Aspiration Biopsy*: An 18 to 27 gauge (most commonly 23-25) needle attached to a 10 cc syringe is used to aspirate (pull out) cells from lesions or masses in various organs of the body by application of negative pressure (suction). FNAC can be done directly on a mass in superficial regions like the neck, thyroid or breast; or it maybe be assisted by ultrasound or CAT scan. FNAC, while poorly developed in the USA, is widely used in Europe and India. Being a skill dependent procedure, the success rate may vary. If performed by pathologist or as team with pathologist-cytotechnologist, the success rate of proper diagnosis is superior. The two countries with the most advanced FNAC services are Sweden (Karolinska hospital performs about 11 thousand annual aspirates), and Slovenia (Institute of Oncology performs about 10 thousand annual aspirates). The highest volumes in USA are encountered at Emory University Hospital in Atlanta GA, and MD Anderson Hostpital in Houston, TX, each contributing no more than 4 thousand aspirates per year.

Equipment

Fine needles are 23 to 27 gauge. Needle diameters and color codes for 23G, 25G and 27G are as follows, respectively: 0,6 mm/ Blue-dark, 0,5 mm/Orange, and 0,4 mm/Grey. If a cell-block preparation is indicated, after obtaining diagnostic cytology smears, a wider gauge needle, up to 18 gauge, may be used.

DEVELOPMENTAL BIOLOGY

Developmental biology is the study of the process by which organisms grow and develop. Modern developmental biology studies the genetic control of cell growth, differentiation and "morphogenesis," which is the process that gives rise to tissues, organs and anatomy. Embryology is a subfield, the study of organisms between the one-cell stage (generally, the zygote) and the end of the embryonic stage, which is not necessarily the beginning of free living. Embryology was originally a more descriptive science until the 20th century. Embryology and developmental biology today deal with the various steps necessary for the correct and complete formation of the body of a living organism.

The related field of evolutionary developmental biology was formed largely in the 1990s and is a synthesis of findings from molecular developmental biology and evolutionary biology which considers the diversity of organismal form in an evolutionary context.

The findings of developmental biology can help to understand developmental malfunctions such as chromosomal aberrations, for example, Down syndrome. An understanding of the specialization of cells during embryogenesis may yield information on how to specialize stem cells to specific tissues and organs, which could lead to the specific cloning of organs for medical purposes. Another biologically important process that occurs during development is apoptosis-programmed cell death or "suicide". For this reason, many developmental models are used to elucidate the physiology and molecular basis of this cellular process. Similarily, a deeper understanding of developmental biology can foster greater progress in the treatment of congenital disorders and diseases, e.g. studying human sex determination can lead to treatment for disorders such as congenital adrenal hyperplasia.

Molecular Mechanisms of Development

During the second half of the 20th century the types of molecules involved in embryonic development were identified. Transcription factors are the key regulators of which genes are expressed in cells. Transcriptional control in the various differentiated cell types allows each type of cell (epithelial, muscle, neuron, etc) to express different amounts of the possible proteins. The transcription factors are regulated by signal transduction pathways that relay signals from outside of cells to the cell nucleus. Signal transduction pathways often involve receptors, receptor ligands and enzymes such as protein kinases. One key class of genes that are differentially regulated by transcription factors in different cell types are genes for cell adhesion proteins. Cell adhesion proteins are among the key regulators of morphogenesis. Functions of these various proteins in development are often elucidated experimentally using gene knockdown techniques in embryos.

Developmental Model Organisms

Often used model organisms in developmental biology include the following:

- Vertebrates
  - o Zebrafish Danio rerio
  - o Medakafish Oryzias latipes
  - o Fugu Takifugu rubripes
  - o Frogs Xenopus laevis
  - o Chicken Gallus gallus
  - o Mouse Mus musculus (Mammalian embryogenesis)
- Invertebrates
  - o Lancelet Branchiostoma lanceolatum
  - o Tunicates
  - o Sea urchin
  - o Round worm Caenorhabditis elegans
  - o Fruit fly Drosophila melanogaster (Drosophila embryogenesis)
- Plants (Plant embryogenesis)
  - o Arabidopsis thaliana
  - o Maize
  - o Snapdragon

Developmental Systems Biology

Computer simulation of multicellular development is a research methodology to understand the function of the very complex processes involved in the development of organisms. This includes simulation of cell signaling, multicell interactions and regulatory genomic networks in development of multicellular structures and processes. Minimal genomes for minimal multicellular organisms may pave the way to understand such complex processes in vivo.

ECOLOGY

Ecology is the scientific study of the distribution and abundance of living organisms and how the distribution and abundance are affected by interactions between the organisms and their environment. The environment of an organism includes both physical properties, which can be described as the sum of local abiotic factors such as insolation (sunlight), climate, and geology, as well as the other organisms that share its habitat. The term oekologie was coined in 1866 by the German biologist Ernst Haeckel the word is derived from the Greek (oikos, "household") and (logos, "study"); therefore "ecology" means the "study of the household of nature". The word "ecology" is often used in common parlance as a synonym for the natural environment or environmentalism. Likewise "ecologic" or "ecological" is often taken in the sense of environmentally friendly.

Scope

Ecology is usually considered a branch of biology, the general science that studies living organisms. Organisms can be studied at many different levels, from proteins and nucleic acids (in biochemistry and molecular biology), to cells (in cellular biology), to individuals (in botany, zoology, and other similar disciplines), and finally at the level of populations, communities, and ecosystems, to the biosphere as a whole; these latter strata are the primary subjects of ecological inquiries. Ecology is a multi-disciplinary science. Because of its focus on the higher levels of the organization of life on earth and on the interrelations between organisms and their environment, ecology draws heavily on many other branches of science, especially geology and geography, meteorology, pedology, genetics, chemistry, and physics. Thus, ecology is considered by some to be a holistic science, one that over-arches older disciplines such as biology which in this view become sub-disciplines contributing to ecological knowledge.

Agriculture, fisheries, forestry, medicine and urban development are among human activities that would fall within Krebs' (1972: 4) explanation of his definition of ecology: "where organisms are found, how many occur there, and why".

As a scientific discipline, ecology does not dictate what is "right" or "wrong". However, ecological knowledge such as the quantification of biodiversity and population dynamics have provided a scientific basis for expressing the aims of environmentalism and evaluating its goals and policies. Additionally, a holistic view of nature is stressed in both ecology and environmentalism. Consider the ways an ecologist might approach studying the life of honeybees:

- The behavioral relationship between individuals of a species is behavioral ecology-for example, the study of the queen bee, and how she relates to the worker bees and the drones.
- The organized activity of a species is community ecology; for example, the activity of bees assures the pollination of flowering plants. Bee hives additionally produce honey which is consumed by still other species, such as bears.
- The relationship between the environment and a species is environmental ecology-for example, the consequences of environmental change on bee activity. Bees may die out due to environmental changes. The environment simultaneously affects and is a consequence of this activity and is thus intertwined with the survival of the species.

Disciplines of Ecology

Ecology is a broad discipline comprised of many sub-disciplines. A common, broad classification, moving from lowest to highest complexity, where complexity is defined as the number of entities and processes in the system under study, is:

- Ecophysiology and Behavioral ecology examine adaptations of the individual to its environment.
- Autecology studies the dynamics of populations of a single species.
- Community ecology (or synecology) focuses on the interactions between species within an ecological community.
- Ecosystem ecology studies the flows of energy and matter through the biotic and abiotic components of ecosystems.
- Landscape ecology examines processes and relationship across multiple ecosystems or very large geographic areas.

Ecology can also be sub-divided according to the species of interest into fields such as animal ecology, plant ecology, insect ecology, and so on. Another frequent method of subdivision is by biome studied, e.g., Arctic ecology (or polar ecology), tropical ecology, desert ecology, etc. The primary technique used for investigation is often used to subdivide the discipline into groups such as chemical ecology, genetic ecology, field ecology, statistical ecology, theoretical ecology, and so forth. Note that these different systems are unrelated and often applied at the same time; one could be a theoretical plant community ecologist, or a polar ecologist interested in animal genetics.

Fundamental Principles of Ecology

Biosphere

The outer layer of the planet Earth can be divided into several compartments: the hydrosphere (or sphere of water), the lithosphere (or sphere of soils and rocks), and the atmosphere (or sphere of the air). The biosphere (or sphere of life), sometimes described as "the fourth envelope", is all living matter on the planet or that portion of the planet occupied by life. It reaches well into the other three spheres, although there are no permanent inhabitants of the atmosphere. Relative to the volume of the Earth, the biosphere is only the very thin surface layer which extends from 11,000 meters below sea level to 15,000 meters above. It is thought that life first developed in the hydrosphere, at shallow depths, in the photic zone. (Although recently a competing theory has emerged, that life originated around hydrothermal vents in the deeper ocean) Multicellular organisms then appeared and colonized benthic zones. Photosynthetic organisms gradually produced the chemically unstable oxygen-rich atmosphere that characterizes our planet. Terrestrial life developed later, after the ozone layer protecting living beings from UV rays formed. Diversification of terrestrial species is thought to be increased by the continents drifting apart, or alternately, colliding. Biodiversity is expressed at the ecological level (ecosystem), population level (intraspecific diversity), species level (specific diversity), and genetic level. Recently technology has allowed the discovery of the deep ocean vent communities. This remarkable ecological system is not dependent on sunlight but bacteria, utilising the chemistry of the hot volcanic vents, are at the base of its food chain.

The biosphere contains great quantities of elements such as carbon, nitrogen and oxygen. Other elements, such as phosphorus, calcium, and potassium, are also essential to life, yet are present in smaller amounts. At the ecosystem and biosphere levels, there is a continual recycling of all these elements, which alternate between the mineral and organic states. While there is a slight input of geothermal energy, the bulk of the functioning of the ecosystem is based on the input of solar energy. Plants and photosynthetic microorganisms convert light into chemical energy by the process of photosynthesis, which creates glucose (a simple sugar) and releases free oxygen. Glucose thus becomes the secondary energy source which drives the ecosystem. Some of this glucose is used directly by other organisms for energy. Other sugar molecules can be converted to other molecules such as amino acids. Plants use some of this sugar, concentrated in nectar to entice pollinators to aid them in reproduction.

Cellular respiration is the process by which organisms (like mammals) break the glucose back down into its constituents, water and carbon dioxide, thus regaining the stored energy the sun originally gave to the plants. The proportion of photosynthetic activity of plants and other photosynthesizers to the respiration of other organisms determines the specific composition of the Earth's atmosphere, particularly its oxygen level. Global air currents mix the atmosphere and maintain nearly the same balance of elements in areas of intense biological activity and areas of slight biological activity.

Water is also exchanged between the hydrosphere, lithosphere, atmosphere and biosphere in regular cycles. The oceans are large tanks, which store water, ensure thermal and climatic stability, as well as the transport of chemical elements thanks to large oceanic currents.

For a better understanding of how the biosphere works, and various dysfunctions related to human activity, American scientists simulated the biosphere in a small-scale model, called Biosphere II.

The Concept of Ecosystem

The first principle of ecology is that each living organism has an ongoing and continual relationship with every other element that makes up its environment. An ecosystem can be defined as any situation where there is interaction between organisms and their environment.

The ecosystem is composed of two entities, the entirety of life, the biocoenosis and the medium that life exists in, the biotope. Within the ecosystem, species are connected by food chains or food webs. Energy from the sun, captured by primary producers via photosynthesis, flows upward through the chain to primary consumers (herbivores), and then to secondary and tertiary consumers (carnivores), before ultimately being lost to the system as waste heat. In the process, matter is incorporated into living organisms, which return their nutrients to the system via decomposition, forming biogeochemical cycles such as the carbon and nitrogen cycles.

The concept of an ecosystem can apply to units of variable size, such as a pond, a field, or a piece of deadwood. A unit of smaller size is called a microecosystem. For example, an ecosystem can be a stone and all the life under it. A mesoecosystem could be a forest, and a macroecosystem a whole ecoregion, with its drainage basin.

The main questions when studying an ecosystem are:

- Whether the colonization of a barren area could be carried out
- Investigation the ecosystem's dynamics and changes
- The methods of which an ecosystem interacts at local, regional and global scale
- Whether the current state is stable
- Investigating the value of an ecosystem and the ways and means that interaction of ecological systems provide benefit to humans, especially in the provision of healthy water.

Ecosystems are often classified by reference to the biotopes concerned. The following ecosystems may be defined:

- As continental ecosystems, such as forest ecosystems, meadow ecosystems such as steppes or savannas), or agro-ecosystems
- As ecosystems of inland waters, such as lentic ecosystems such as lakes or ponds; or lotic ecosystems such as rivers
- As oceanic ecosystems.

Another classification can be done by reference to its communities, such as in the case of an human ecosystem.

Dynamics and Stability

Ecological factors which affect dynamic change in a population or species in a given ecology or environment are usually divided into two groups: abiotic and biotic.

Abiotic factors are geological, geographical, hydrological and climatological parameters. A biotope is an environmentally uniform region characterized by a particular set of abiotic ecological factors. Specific abiotic factors include:

- Water, which is at the same time an essential element to life and a milieu

- Air, which provides oxygen, nitrogen, and carbon dioxide to living species and allows the dissemination of pollen and spores
- Soil, at the same time source of nutriment and physical support
  - o Soil pH, salinity, nitrogen and phosphorus content, ability to retain water, and density are all influential
- Temperature, which should not exceed certain extremes, even if tolerance to heat is significant for some species
- Light, which provides energy to the ecosystem through photosynthesis
- Natural disasters can also be considered abiotic

Biocenose, or community, is a group of populations of plants, animals, micro-organisms. Each population is the result of procreations between individuals of same species and cohabitation in a given place and for a given time. When a population consists of an insufficient number of individuals, that population is threatened with extinction; the extinction of a species can approach when all biocenoses composed of individuals of the species are in decline. In small populations, consanguinity (inbreeding) can result in reduced genetic diversity that can further weaken the biocenose.

Biotic ecological factors also influence biocenose viability; these factors are considered as either intraspecific and interspecific relations.

Intraspecific relations are those which are established between individuals of the same species, forming a population. They are relations of co-operation or competition, with division of the territory, and sometimes organization in hierarchical societies.

Interspecific relations-interactions between different species-are numerous, and usually described according to their beneficial, detrimental or neutral effect (for example, mutualism (relation ++) or competition (relation--). The most significant relation is the relation of predation (to eat or to be eaten), which leads to the essential concepts in ecology of food chains (for example, the grass is consumed by the herbivore, itself consumed by a carnivore, itself consumed by a carnivore of larger size).

A high predator to prey ratio can have a negative influence on both the predator and prey biocenoses in that low availability of food and high death rate prior to sexual maturity can decrease (or prevent the increase of) populations of each, respectively. Selective hunting of species by humans which leads to population decline is one example of a high predator to prey ratio in action. Other interspecific relations include parasitism, infectious disease and competition for limiting resources, which can occur when two species share the same ecological niche.

The existing interactions between the various living beings go along with a permanent mixing of mineral and organic substances, absorbed by organisms for their growth, their maintenance and their reproduction, to be finally rejected as waste. These permanent recyclings of the elements (in particular carbon, oxygen and nitrogen) as well as the water are called biogeochemical cycles.

They guarantee a durable stability of the biosphere (at least when unchecked human influence and extreme weather or geological phenomena are left aside). This self-regulation, supported by negative feedback controls, ensures the perenniality of the ecosystems. It is shown by the very stable concentrations of most elements of each compartment. This is referred to as homeostasis. The ecosystem also tends to evolve to a state of ideal balance, reached after a succession of events, the climax (for example a pond can become a peat bog).

Spatial Relationships and Subdivisions of Land

Ecosystems are not isolated from each other, but are interrelated. For example, water may circulate between ecosystems by the means of a river or ocean current. Water itself, as a liquid medium, even defines ecosystems. Some species, such as salmon or freshwater eels move between marine systems and fresh-water systems. These relationships between the ecosystems lead to the concept of a biome. A biome is a homogeneous ecological formation that exists over a large region as tundra or steppes. The biosphere comprises all of the Earth's biomes--the entirety of places where life is possible--from the highest mountains to the depths of the oceans. Biomes correspond rather well to subdivisions distributed along the latitudes, from the equator towards the poles, with differences based on to the physical environment (for example, oceans or mountain ranges) and to the climate. Their variation is generally related to the distribution of species according to their ability to tolerate temperature and/or dryness. For example, one may find photosynthetic algae only in the photic part of the ocean (where light penetrates), while conifers are mostly found in mountains.

Though this is a simplification of more complicated scheme, latitude and altitude approximate a good representation of the distribution of biodiversity within the biosphere. Very generally, the richness of biodiversity (as well for animal than plant species) is decreasing most rapidly near the equator and less rapidly as one approaches the poles.

The biosphere may also be divided into ecozone, which are very well defined today and primarily follow the continental borders. The ecozones are themselves divided into ecoregions, though there is not agreement on their limits.

Ecosystem Productivity

In an ecosystem, the connections between species are generally related to food and their role in the food chain. There are three categories of organisms:

- Producers--usually plants which are capable of photosynthesis but could be other organisms such as bacteria around ocean vents that are capable of chemosynthesis.
- Consumers--animals, which can be primary consumers (herbivorous), or secondary or tertiary consumers (carnivorous).
- Decomposers--bacteria, mushrooms which degrade organic matter of all categories, and restore minerals to the environment.

These relations form sequences, in which each individual consumes the preceding one and is consumed by the one following, in what are called food chains or food network. In a food network, there will be fewer organisms at each level as one follows the links of the network up the chain. These concepts lead to the idea of biomass (the total living matter in a given place), of primary productivity (the increase in the mass of plants during a given time) and of secondary productivity (the living matter produced by consumers and the decomposers in a given time).

These two last ideas are key, since they make it possible to evaluate the load capacity--the number of organisms which can be supported by a given ecosystem. In any food network, the energy contained in the level of the producers is not completely transferred to the consumers. And the higher one goes up the chain, the more energy and resources is lost and consumed.

Thus, from an energy-and environmental-point of view, it is more efficient for humans to be primary consumers (to subsist from vegetables, grains, legumes, fruit, cotton, etc.) than as secondary consumers (from eating herbivores, omnivores, or their products, such as milk, chickens, cattle, sheep, etc.) and still more so than as a tertiary consumer (from consuming carnivores, omnivores, or their products, such as fur, pigs, snakes, alligators, etc.). An ecosystem(s) is unstable when the load capacity is overrun and is especially unstable when a population doesn't have an ecological niche and overconsumers.

The productivity of ecosystems is sometimes estimated by comparing three types of land-based ecosystems and the total of aquatic ecosystems:

- The forests (1/3 of the Earth's land area) contain dense biomasses and are very productive. The total production of the world's forests corresponds to half of the primary production.
- Savannas, meadows, and marshes (1/3 of the Earth's land area) contain less dense biomasses, but are productive. These ecosystems represent the major part of what humans depend on for food.

- Extreme ecosystems in the areas with more extreme climates--deserts and semi-deserts, tundra, alpine meadows, and steppes--(1/3 of the Earth's land area) have very sparse biomasses and low productivity.
- Finally, the marine and fresh water ecosystems (3/4 of Earth's surface) contain very sparse biomasses (apart from the coastal zones).

Humanity's actions over the last few centuries have seriously reduced the amount of the Earth covered by forests (deforestation), and have increased agro-ecosystems (agriculture). In recent decades, an increase in the areas occupied by extreme ecosystems has occurred (desertification).

Ecological Crisis

Generally, an ecological crisis occurs with the loss of adaptive capacity when the resilience of an environment or of a species or a population evolves in a way unfavourable to coping with perturbations that interfere with that ecosystem, landscape or species survival. It may be that the environment quality degrades compared to the species needs, after a change in an abiotic ecological factor (for example, an increase of temperature, less significant rainfalls).

It may be that the environment becomes unfavourable for the survival of a species (or a population) due to an increased pressure of predation (for example overfishing). Lastly, it may be that the situation becomes unfavourable to the quality of life of the species (or the population) due to a rise in the number of individuals (overpopulation).

Ecological crises may be more or less brutal (occurring within a few months or taking as long as a few million years). They can also be of natural or anthropic origin. They may relate to one unique species or to many species. Lastly, an ecological crisis may be local (as an oil spill) or global (a rise in the sea level due to global warming).

According to its degree of endemism, a local crisis will have more or less significant consequences, from the death of many individuals to the total extinction of a species. Whatever its origin, disappearance of one or several species often will involve a rupture in the food chain, further impacting the survival of other species.

In the case of a global crisis, the consequences can be much more significant; some extinction events showed the disappearance of more than 90% of existing species at that time. However, it should be noted that the disappearance of certain species, such as the dinosaurs, by freeing an ecological niche, allowed the development and the diversification of the mammals. An ecological crisis thus paradoxically favored biodiversity.

Sometimes, an ecological crisis can be a specific and reversible phenomenon at the ecosystem scale. But more generally, the crises impact will last. Indeed, it rather is a connected series of events, that occur till a final point. From this stage, no return to the previous stable state is possible, and a new stable state will be set up gradually.

Lastly, if an ecological crisis can cause extinction, it can also more simply reduce the quality of life of the remaining individuals. Thus, even if the diversity of the human population is sometimes considered threatened, few people envision human disappearance at short span. However, epidemic diseases, famines, impact on health of reduction of air quality, food crises, reduction of living space, accumulation of toxic or non degradable wastes, threats on keystone species (great apes, panda, whales) are also factors influencing the well-being of people.

During the past decades, this increasing responsibility of humanity in some ecological crises has been clearly observed. Due to the increases in technology and a rapidly increasing population, humans have more influence on their own environment than any other ecosystem engineer.

Some usually quoted examples as ecological crises are:

- *Permian*: Triassic extinction event 250 million of years ago
- *Cretaceous*: Tertiary extinction event 65 million years ago
- Global warming related to the Greenhouse effect. Warming could involve flooding of the Asian deltas, multiplication

of extreme weather phenomena and changes in the nature and quantity of the food resources.

- Ozone layer hole issue
- Deforestation and desertification, with disappearance of many species.
- The nuclear meltdown at Chernobyl in 1986 caused the death of many people and animals from cancer, and caused mutations in a large number of animals and people. The area around the plant is now abandoned by humans because of the large amount of radiation generated by the meltdown. Twenty years after the accident, the animals have returned.

2

BIOLOGICAL GENETICS

Genetics (from the Greek *genno* = give birth) is the science of genes, heredity, and the variation of organisms. The word "genetics" was first suggested to describe the study of inheritance and the science of variation by the prominent British scientist William Bateson in a personal letter to Adam Sedgwick, dated April 18, 1905. Bateson first used the term "genetics" publicly at the Third International Conference on Plant Hybridization (London, England) in 1906.

Heredity and variations form the basis of genetics. Humans applied knowledge of genetics in prehistory with the domestication and breeding of plants and animals. In modern research, genetics provides important tools for the investigation of the function of a particular gene, e.g., analysis of genetic interactions. Within organisms, genetic information generally is carried in chromosomes, where it is represented in the chemical structure of particular DNA (deoxyribonucleic acid) molecules.

Genes encode the information necessary for synthesizing the amino-acid sequences in proteins, which in turn play a large role in determining the final phenotype, or physical appearance, of the organism. In diploid organisms, a dominant allele on one chromosome will mask the expression of a recessive gene on the other.

The phrase to code for is often used to mean a gene contains the instructions about how to build a particular protein, as in the gene codes for the protein. The "one gene, one protein" concept is now known to be simplistic. For example, a single gene may produce multiple products, depending on how its transcription is regulated. Genes code for the nucleotide sequences in mRNA, tRNA and rRNA, required for protein synthesis.

Genetics determines much (but not all) of the appearance of organisms, including humans, and possibly how they act. Environmental differences and random factors also play a part. Monozygotic ("identical") twins, a clone resulting from the early splitting of an embryo, have the same DNA, but different personalities and fingerprints. Genetically-identical plants grown in colder climates incorporate shorter and less-saturated fatty acids to avoid stiffness.

ORIGIN AND HISTORY

History of Genetics

The history of genetics is generally held to have started with the work Austrian monk, Gregor Mendel; his work on pea plants, published in 1866, described what came to be known as Mendelian inheritance. In the centuries before-and for several decades after-Mendel's work, an wide variety of theories of heredity proliferated. 1900 marked the "rediscovery of Mendel" by Hugo de Vries, Carl Correns and Erich von Tschermak, and by 1915 the basic principles of Mendelian genetics had been applied to a wide variety of organisms-most notably the fruit fly Drosophila melanogaster. Led by Thomas Hunt Morgan and his fellow "drosophilists", geneticists developed the Mendelian-chromosome theory of heredity, which was widely accepted by 1925. Alongside experimental work, mathematicians developed the statistical framework of population genetics, bring genetical explanations into the study of evolution.

With the basic patterns of genetic inheritance established, many biologists turned to investigations of the physical nature of the gene. In the 1940s and early 1950s, experiments pointed to DNA as the portion of chromosomes (and perhaps other nucleoproteins) that held genes. A focus on new model organisms such as viruses and bacteria, along with the discovery of the double helical structure of DNA in 1953, marked the transition to the era of molecular genetics. In the following years, chemists developed techniques for sequencing both nucleic acids and proteins, while others worked out the relationship between the two forms of biological molecules: the genetic code. The regulation of gene expression became a central problem in the 1960s; by the 1970s gene expression could be controlled and manipulated through genetic engineering. In the last decades of the 20th century, many biologists focused on large-scale genetics projects, sequencing entire genomes.

Pre-Mendelian Ideas on Heredity

Ancient Theories

The most influential early theories of heredity were that of Hippocrates and Aristotle. Hippocrates' theory (possibly based on the teachings of Anaxagoras) was similar to Darwin's later ideas on pangenesis, involving heredity material that collects from throughout the body. Aristotle suggested instead that the (nonphysical) form-giving principle of an organism was transmitted through semen, determining the shape of the female's menstrual blood through an organism's early development. For both Hippocrates and Aristotle-and nearly all Western scholars through the late 19th century-the inheritance of acquired characters was a supposedly well-established fact that any adequate theory of heredity had to explain. At the same time, individual species were taken to have a fixed essence; such inherited changes were merely superficial.

Plant Systematics and Hybridization

In the 18th century, with increased knowledge of plant and animal diversity and the accompanying increased focus on taxonomy, new ideas about heredity began to appear. Linnaeus and others (among them Joseph Gottlieb Kölreuter, Carl Friedrich von Gärtner, and Charles Naudin) conducted extensive experiments with hybridization, especially species hybrids. Species hybridizers described a wide variety of inheritance phenomena, include hybrid sterility and the high variability of back-crosses.

Plant breeders were also developing an array of stable varieties in many important plant species. In the early 19th century, Augustin Sageret established the concept of dominance, recognizing that when some plant varieties are crossed, certain characters (present in one parent) usually appear in the offspring; he also found that some ancestral characters found in neither parent may appear in offspring. However, plant breeders made little attempt to develop a theoretical foundation for their work or to integrate their knowledge with work in physiology and natural history.

Mendel

In breeding experiments between 1856 and 1865, Gregor Mendel first traced inheritance patterns of certain traits in pea plants and showed that they obeyed simple statistical rules. Although not all features show these patterns of Mendelian inheritance, his work acted as a proof that application of statistics to inheritance could be highly useful. Since that time many more complex forms of inheritance have been demonstrated.

From his statistical analysis Mendel defined a concept that he described as an allele, which was the fundamental unit of heredity. The term allele as Mendel used it is nearly synonymous with the term gene, and now means a specific variant of a particular gene.

Mendel's work was published in 1866 as Experiments on Plant Hybridization in the Proceedings of the Natural History Society of Brünn, following two lectures he gave on the work in early 1865.

Post-Mendel, Pre-re-discovery

Mendel's work was published in a relatively obscure scientific journal, and it was not given any attention in the scientific community. Instead, discussions about modes of heredity were galvanized by Darwin's theory of evolution by natural selection, in which mechanisms of non-Lamarckian heredity seemed to be required. Darwin's own theory of heredity, pangenesis, did not meet with any large degree of acceptance. A more mathematical version of pangenesis, one which dropped much of Darwin's

Lamarckian holdovers, was developed as the "biometrical" school of heredity by Darwin's cousin, Francis Galton. Under Galton and his successor Karl Pearson, the biometrical school attempted to build statistical models for heredity and evolution, with some limited but real success, though the exact methods of heredity were unknown and largely unquestioned.

GREAT CHAIN OF BEING

The great chain of being or scala naturæ is a classical and western medieval conception of the order of the universe, whose chief characteristic is a strict hierarchical system.

It is a conception of the world's structure that was accepted, and unquestioned, by most educated men from the time of Lucretius until the Copernican and Darwinian revolution and the ultimate flowering of the Renaissance. The chain of being is composed of a great number of hierarchal links, from the most basic and foundational elements up through the very highest perfection, in other words, God, or the Prime Mover.

God, and beneath him the angels, both existing wholly in spirit form, sit at the top of the ladder. Earthly flesh is fallible and ever-changing: mutable. Spirit, however, is unchanging and permanent. This sense of permanence is crucial to understanding this conception of reality. One does not abandon one's place in the chain; it is not only unthinkable, but generally impossible. (One exception might be in the realm of alchemy, where alchemists attempted to transmute base elements, such as lead, into higher elements, either silver, or more often, gold-the highest element.) In the natural order, earth (rock is at the bottom of the chain; these elements possess only the attribute of existence. Moving on up the chain, each succeeding link contains the positive attributes of the previous link, and adds (at least) one other. Rocks, as above, possess only existence; the next link up, plants, possess life and existence. Beasts add not only motion, but appetite as well.

Man is a special instance in this conception. He is both mortal flesh, as those below him, and also spirit. In this dichotomy, the struggle between flesh and spirit becomes a moral one. The way of the spirit is higher, more noble; it brings one closer to God. The desires of the flesh drag one down. The Christian fall of Lucifer is especially terrible, because that angel is wholly spirit, who yet defies God, the ultimate perfection.

Diagram

The purpose of the great chain of being was to assign a place for everything in the universe, classified into either:

- holy
- animal
- vegetable
- mineral

Example: If one were to examine only the earthly inhabitants, and their place in the chain, this is what would be found:

- God
- angels
- humans
- birds
- fish
- mammals
- plants
- rocks

The concept of "animal, vegetable, and mineral" still survives and is a feature of many Western philosophies and cultures to this day, and importantly made up the "three kingdoms" of Carolus Linnaeus's taxonomic system.

Other Subdivisions

Each link in the chain might be further divided into its component parts. In medieval secular society, for example, the king is at the top, followed by the aristocratic lords, and then the peasants below them. Solidifying the king's position at the top of humanity's social order is the doctrine of the divine right of kings. In the family, the father is head of the household; below him, his wife; below her, their children. The children might be subdivided so that the males are one link above the females. Modern western culture maintains some of these divisions. Just as Milton's Paradise Lost ranked the angels (c.f. Pseudo-Dionysius the Areopagite's ranking of angels), so Christian culture conceives of angels in orders of archangels, seraphim, and cherubim, among others. The lion is still king of light.

Amongst animals, subdivisions are equally apparent. At the top of the animals are wild beasts, who were seen as superior as they defied training and domestication. Below them are domestic animals, further sub-divided so that useful animals (such as dogs and horses) are higher than docile creature such as sheep. Birds are also sub-divided, with eagles above pigeons, for example. Fish come below birds, and are sub-divided between actual fish and other sea creatures which are abhorred in the Bible. Below them come insects, with useful insects such spiders and bees, and attractive creatures such as ladybirds and dragonflies at the top, and unpleasant insects such as flies and beetles at the bottom. At the very bottom of the animal sector are snakes, which are relegated to this position as punishment for the serpent's actions in the Garden of Eden.

Below animals comes the division for plants, which is further sub-divided. Trees are at the top, with useful trees such as oaks at the top, and the traditionally demonic yew tree at the bottom. Food-producing plants such as cereals and vegetables, further sub-divided.

At the very bottom of the chain are minerals. At the top of this section are metals (further sub-divided, with gold at the top), rocks (with granite and marble at the top), soil (sub-divided between nutrient-rich soil and low-quality types), sand, grit, dust, and dirt. A reference to the Great Chain of Being that survives in today's English language is the insult that one is "lower than dirt," which refers to dirt's place at the bottom of the Chain.

The central concept of the chain of being is that everything imaginable fits into it somewhere, giving order and meaning to the universe.

HOLINESS

Holiness is the state of being holy, that is, set apart for the worship or service of God or gods. It is most usually ascribed to people, but can be and often is ascribed to objects, times, or places. The word holy is related to the word whole.

The origin of the word comes from the 11th Century Old High German hulis and Old English holegn meaning Holly as in Holly Tree, considered a sacred plant to both pre-Christian Celtic and Roman worship. The word hulis originates from an even older proto-Germanic word khuli a shortened derivation of the ancient Gaelic cuilieann both meaning Holly. The distinction of the word holy appeared around the 13th Century with the Old English word halig derived from hal meaning health, happiness and wholeness. As "wholeness", holiness may be taken to indicate a state of religious completeness or perfection.

The concept is found in almost all religions, especially those with a degree of formal organization, so that there is scope for special people, places etc to be devoted to worship.

In non-specialist contexts, the term "holy" is used in a more general way, to refer to someone or something that is associated with a divine power, such as water used for Baptism.

Holiness in Judaism

The Judaic tradition conceives 'holiness' in various ways, but most distinctive and central to the Torah is the priestly system, legislation for which comprises the majority of the law delivered at Sinai. It essentially involves the division of time and space into the spheres of the divine and the common. The word used in Leviticus for this separation, hivdil, is that used for the process of Creation in Genesis. Creation is a matter of proper ordering, which involves not only land and water, light and darkness, but also holy and profane, clean and unclean. It is the role of the priesthood, and Israel as God's priestly nation, to maintain this order in accordance with the guidelines set out at Sinai. The safety and stability of each individual, the nation, and ultimately the whole world, depends on it.

The division seems to be understood as a gradual one, with different behaviour appropriate to the merely 'holy' and the 'most holy'. The Mishnah therefore lists concentric circles of holiness: Holy of Holies; sanctuary; vestibule; court of priests; court of Israelites; court of women; temple mount; Jerusalem walls; all the walled cities of Israel; and the borders of the holy land. Distinctions are made as to who and what are permitted in each area. Likewise the calendar is divided so that the eve of the Sabbath to the end of the day is holy time, and certain feasts, such as the Day of Atonement, are most holy. Both holy time and holy space are rooted in Creation, with the Sabbath as its culmination, and the Garden traditionally on the site of the temple.

So holiness denotes the sphere of the divine, which is to be set apart, and is manifest in power particularly when its separation is not properly maintained. There are various stories in the Hebrew Bible of disease and destruction resulting from improper contact with or handling of holy things such as the Ark. This dynamic power is divine, and so the holy is very much associated with the divine Presence. The relationship between holiness and Presence is unclear: holiness seems to be a precondition for the manifestation of the Presence, but is not to be equated with it. In practical terms, holiness can be measured and managed by priestly legislation, while Presence is entirely dependent upon God's action. The priestly conception of holiness expresses the distinctively scriptural perception of God as both transcendent (utterly separate) and powerfully immanent in His relationship with the world.

A person ascribed with holiness in Judaism are known as a Tzadik.

Holiness in Christianity

Holiness in Catholicism

Catholicism has adopted much of the Jewish vision of the world in terms of holiness, with certain behaviour appropriate to certain places and times. The calendar gives shape to Catholic practice, which tends to focus on the Eucharist, in which the Real Presence of Christ is manifested. Many features of the Jewish temple are imitated in churches, such as the altar, bread, lamp, incense, font, etc, to emphasise the extreme holiness of the Eucharistic elements, which are reserved in a tabernacle. In extension of this focus on the Sacrament as holy, many objects in Catholicism are also considered holy. They are called sacramentals and are usually blessed by a priest. Such items include rosaries, crucifixes, medals, and statues of Jesus, angels and saints (Virgin Mary).

People in a state of grace are also considered holy in Catholicism. A central notion of Catholicism as articulated in contemporary theology is the "[personal] call to holiness," considered as a vocation shared by every Christian believer. Profound personal holiness has traditionally also been seen as a focus for the kind of contagious holiness primarily associated with the Sacrament. So the cult of saints in Catholicism is not only the acclamation of their piety or morality, but also reverence for the tangible holiness that flows from their proximity to the divine. Hence the places where saints lived, died, performed miracles, or received visions frequently become sites of pilgrimage, and notable objects surviving a saint (including the body or parts of it) are considered relics. The holiness of such places or objects, resulting from contact with a deeply holy person, is often connected with the miraculous long after the death of the saint.

Holiness in Protestantism

The Protestant Reformation stood in opposition to the beliefs of tangible holiness in the Catholic Church and rejected most of its teachings regarding devotional practice, language and imagery. The early Reformers, who were often scholars of ancient Greek and also borrowed from Jewish scholarship, recognized that holiness is an attribute of God, and holiness is always part of the presence of God. Yet they also recognized that practical holiness was the evidence of the presence of God in the converted believer. Martin Luther, viewed God's grace (and therefore God's holiness), as an invasion of the life. Actions that demonstrated holiness would spring up, not premeditated, as the believer focused more and more on his or her relationship with Christ. This was the life of faith, according to Luther, a life in which one recognizes that the sin nature never departs, yet grace invades and draws the person after Christ.

Calvin, on the other hand, formulated a practical system of holiness that even tied in with culture and social justice. All unholy actions, Calvin reasoned, resulted in suffering. Thus he proved out to the city fathers of Geneva that dancing and other social vices always ended with the wealthy oppressing the poor. A holy life, in his outlook, was pietistic and simple, a life that shunned extravagance, excess, and vanity. On a personal level, Calvin believed that suffering would be a manifestation of taking on the Cross of Christ, but suffering was also part of the process of holiness.

He expected that all Christians would suffer in this life, not as punishment, but rather as participation in union with Christ, who suffered for them. And yet, socially, Calvin argued that a holy society would end up as a gentle, kindly society (except to criminals), where the poor would be protected from the abuses of the wealthy, the lawyers, and others who normally preyed upon them. In Protestantism, especially in American branches of Protestantism of the more Pentecostal variety, holiness has acquired the secondary meaning of the reshaping of a person through spiritual rebirth. The term owes its origin to John Wesley's concept of "scriptural holiness" or Christian perfection.

The Holiness movement began within Methodism in the United States, among those who thought the church had lost the zeal and emphasis on personal holiness of Wesley's day. In the latter part of the 19th century revival meetings were held, attended by thousands. In Vineland, N.J in 1867 a camp meeting was begun and the National Holiness Camp Meeting Association, which went on to establish many holiness camp meetings across the nation. Some adherents to the movement remained within their denominations; others founded new denominations, such as the Free Methodist Church, the Church of the Nazarene, and the Church of God (Anderson). Within a generation another movement, the Pentecostal movement was born, drawing heavily from the Holiness Movement. Around the middle of the 20th century, the Conservative Holiness Movement was born-a conservative offshoot of the Holiness movement.

In the contemporary Holiness movement, the idea that holiness is relational is growing. In this thought, the core notion of holiness is love. Other notions of holiness, such as purity, being set apart, perfection, keeping rules, and total commitment, are seen as contributory notions of holiness. These contributory notions find their ultimate legitimacy one when love is at their core (Thomas Jay Oord and Michael Lodahl).

Holiness in Buddhism

In Theravada Buddhism one finds the designation of 'noble person' or ariyapuggala (Pali). The Buddha described four grades of such person depending on their level of purity. This purity is measured by which of the ten fetters (samyojana) have been removed from the personality or psyche of that person, These persons are called (in order of increasing sanctity) Sotapanna, Sakadagami, Anagami and Arahant. The latter term designates an enlightened human being and is sometimes rendered into English as the Holy One.

ANIMAL

Animals are a major group of organisms, classified as the kingdom Animalia or Metazoa. In general they are multicellular, capable of locomotion at some stage in their life cycle, responsive to their environment, and feed by consuming other organisms or parts of them. Their body plan becomes fixed as they develop, usually early on in their development as embryos, although some undergo a process of metamorphosis later on.

The word "animal" comes from the Latin word animal, of which animalia is the plural, and is derived from anima, meaning vital breath or soul. In everyday colloquial usage, the word usually refers to non-human animals. The biological definition of the word refers to all members of the Kingdom Animalia. Therefore, when the word "animal" is used in a biological context, humans are included.

Characteristics

Animals have several characteristics that set them apart from other living things. Animals are eukaryotic and usually multicellular, which separates them from bacteria and most protists. They are heterotrophic, generally digesting food in an internal chamber, which separates them from plants and algae. They are also distinguished from plants, algae, and fungi by lacking cell walls.

Structure

With a few exceptions, most notably the sponges (Phylum Porifera), animals have bodies differentiated into separate tissues. These include muscles, which are able to contract and control locomotion, and nerve tissue, which sends and processes signals. There is also typically an internal digestive chamber, with one or two openings. Animals with this sort of organization are called metazoans, or eumetazoans when the former is used for animals in general. All animals have eukaryotic cells, surrounded by a characteristic extracellular matrix composed of collagen and elastic glycoproteins. This may be calcified to form structures like shells, bones, and spicules. During development it forms a relatively flexible framework upon which cells can move about and be reorganized, making complex structures possible. In contrast, other multicellular organisms like plants and fungi have cells held in place by cell walls, and so develop by progressive growth. Also, unique to animal cells are the following intercellular junctions: tight junctions, gap junctions, and desmosomes.

Reproduction and Development

Nearly all animals undergo some form of sexual reproduction. Adults are diploid or polyploid. They have a few specialized reproductive cells, which undergo meiosis to produce smaller motile spermatozoa or larger non-motile ova. These fuse to form zygotes, which develop into new individuals. Many animals are also capable of asexual reproduction. This may take place through parthenogenesis, where fertile eggs are produced without mating, or in some cases through fragmentation.

A zygote initially develops into a hollow sphere, called a blastula, which undergoes rearrangement and differentiation. In sponges, blastula larvae swim to a new location and develop into a new sponge. In most other groups, the blastula undergoes more complicated rearrangement. It first invaginates to form a gastrula with a digestive chamber, and two separate germ layers-an external ectoderm and an internal endoderm. In most cases, a mesoderm also develops between them. These germ layers then differentiate to form tissues and organs.

Most animals grow by indirectly using the energy of sunlight. Plants use this energy to convert sunlight into simple sugars using a process known as photosynthesis. Starting with the molecules carbon dioxide (CO_2) and water (H_2O), photosynthesis converts the energy of sunlight into chemical energy stored in the bonds of glucose ($C_6H_{12}O_6$) and releases oxygen (O_2). These sugars are then used as the building blocks which allow the plant to grow. When animals eat these plants (or eat other animals which have eaten plants), the sugars produced by the plant are used by the animal. They are either used directly to help the animal grow, or broken down, releasing stored solar energy, and giving the animal the energy required for motion. This process is known as glycolysis.

Animals who live close to hydrothermal vents and cold seeps on the ocean floor are not dependent on the energy of sunlight. Instead, chemosynthetic archaea and eubacteria form the base of the food chain.

Origin and Fossil Record

Animals are generally considered to have evolved from flagellate protozoa. Their closest living relatives are the choanoflagellates, collared flagellates that have the same structure as certain sponge cells do. Molecular studies place them in a supergroup called the opisthokonts, which also include the fungi and a few small parasitic protists. The name comes from the posterior location of the flagellum in motile cells, such as most animal spermatozoa, whereas other eukaryotes tend to have anterior flagella.

The first fossils that might represent animals appear towards the end of the Precambrian, around 575 million years ago, and are known as the Ediacaran or Vendian biota. These are difficult to relate to later fossils, however. Some may represent precursors of modern phyla, but they may be separate groups, and it is possible they are not really animals at all.

Aside from them, most known animal phyla make a more or less simultaneous appearance during the Cambrian period, about 542 million years ago. It is still disputed whether this event, called the Cambrian explosion, represents a rapid divergence between different groups or a change in conditions that made fossilization possible.

Groups of Animals

The sponges (Porifera) diverged from other animals early. As mentioned above, they lack the complex organization found in most other phyla. Their cells are differentiated, but not organized into distinct tissues. Sponges are sessile and typically feed by drawing in water through pores. Archaeocyatha, which have fused skeletons, may represent sponges or a separate phylum.

Among the eumetazoan phyla, two are radially symmetric and have digestive chambers with a single opening, which serves as both the mouth and the anus. These are the Cnidaria, which include sea anemones, corals, and jellyfish, and the Ctenophora or comb jellies. Both have distinct tissues, but they are not organized into organs. There are only two main germ layers, the ectoderm and endoderm, with only scattered cells between them. As such, these animals are sometimes called diploblastic. The tiny Placozoans are similar, but they do not have a permanent digestive chamber.

The remaining animals form a monophyletic group called the Bilateria. For the most part, they are bilaterally symmetric, and often have a specialized head with feeding and sensory organs. The body is triploblastic, i.e. all three germ layers are well-developed, and tissues form distinct organs. The digestive chamber has two openings, a mouth and an anus, and there is also an internal body cavity called a coelom or pseudocoelom. There are exceptions to each of these characteristics, however-for instance adult echinoderms are radially symmetric, and certain parasitic worms have extremely simplified body structures.

Genetic studies have considerably changed our understanding of the relationships within the Bilateria. Most appear to belong to four major lineages:

1. Deuterostomes
2. Ecdysozoa
3. Platyzoa
4. Lophotrochozoa

In addition to these, there are a few small groups of bilaterians with relatively similar structure that appear to have diverged before these major groups. These include the Acoelomorpha, Rhombozoa, and Orthonectida. The Myxozoa, single-celled parasites that were originally considered Protozoa, are now believed to have developed from the Bilateria as well.

Deuterostomes

Deuterostomes differ from the other Bilateria, called protostomes, in several ways. In both cases there is a complete digestive tract. However, in protostomes the initial opening (the archenteron) develops into the mouth, and an anus forms separately. In deuterostomes this is reversed. In most protostomes cells simply fill in the interior of the gastrula to form the mesoderm, called schizocoelous development, but in deuterostomes it forms through invagination of the endoderm, called enterocoelic pouching. Deuterostomes also have a dorsal, rather than a ventral, nerve chord and their embryos undergo different cleavage.

All this suggests the deuterostomes and protostomes are separate, monophyletic lineages. The main phyla of deuterostomes are the Echinodermata and Chordata. The former are radially symmetric and exclusively marine, such as starfish, sea urchins, and sea cucumbers. The latter are dominated by the vertebrates, animals with backbones. These include fish, amphibians, reptiles, birds, and mammals. In addition to these, the deuterostomes also include the Hemichordata or acorn worms. Although they are not especially prominent today, the important fossil graptolites may belong to this group. The Chaetognatha or arrow worms may also be deuterostomes, but more recent studies suggest protostome affinities.

Ecdysozoa

The Ecdysozoa are protostomes, named after the common trait of growth by moulting or ecdysis. The largest animal phylum belongs here, the Arthropoda, including insects, spiders, crabs, and their kin. All these organisms have a body divided into repeating segments, typically with paired appendages. Two smaller phyla, the Onychophora and Tardigrada, are close relatives of the arthropods and share these traits.

The ecdysozoans also include the Nematoda or roundworms, the second largest animal phylum. Roundworms are typically microscopic, and occur in nearly every environment where there is water. A number are important parasites. Smaller phyla related to them are the Nematomorpha or horsehair worms, which are invisible to the unaided eye, and the Kinorhyncha, Priapulida, and Loricifera. These groups have a reduced coelom, called a pseudocoelom.

The remaining two groups of protostomes are sometimes grouped together as the Spiralia, since in both embryos develop with spiral cleavage.

Platyzoa

The Platyzoa include the phylum Platyhelminthes, the flatworms. These were originally considered some of the most primitive Bilateria, but it now appears they developed from more complex ancestors. A number of parasites are included in this group, such as the flukes and tapeworms. Flatworms lack a coelom, as do their closest relatives, the microscopic Gastrotricha.

The other platyzoan phyla are microscopic and pseudocoelomate. The most prominent are the Rotifera or rotifers, which are common in aqueous environments. They also include the Acanthocephala or spiny-headed worms, the Gnathostomulida, Micrognathozoa, and possibly the Cycliophora. These groups share the presence of complex jaws, from which they are called the Gnathifera.

Lophotrochozoa

The Lophotrochozoa include two of the most successful animal phyla, the Mollusca and Annelida. The former includes animals such as snails, clams, and squids, and the latter comprises the segmented worms, such as earthworms and leeches. These two groups have long been considered close relatives because of the common presence of trochophore larvae, but the annelids were considered closer to the arthropods, because they are both segmented. Now this is generally considered convergent evolution, owing to many morphological and genetic differences between the two phyla.

The Lophotrochozoa also include the Nemertea or ribbon worms, the Sipuncula, and several phyla that have a fan of cilia around the mouth, called a lophophore. These were traditionally grouped together as the lophophorates, but it now appears they are paraphyletic, some closer to the Nemertea and some to the Mollusca and Annelida. They include the Brachiopoda or lamp shells, which are prominent in the fossil record, the Entoprocta, the Phoronida, and possibly the Bryozoa or moss animals.

History of Classification

Aristotle divided the living world between animals and plants, and this was followed by Carolus Linnaeus in the first hierarchical classification. Since then biologists have begun emphasizing evolutionary relationships, and so these groups have been restricted somewhat. For instance, microscopic protozoa were originally considered animals because they move, but are now treated separately.

In Linnaeus' original scheme, the animals were one of three kingdoms, divided into the classes of Vermes, Insecta, Pisces, Amphibia, Aves, and Mammalia. Since then the last four have all been subsumed into a single phylum, the Chordata, whereas the various other forms have been separated out. The above lists represent our current understanding of the group, though there is some variation from source to source.

In his paper "Versuche über Pflanzenhybriden" ("Experiments in Plant Hybridization"), presented in 1865 to the Brunn Natural History Society, Gregor Mendel traced the inheritance patterns of certain traits in pea plants and showed that they could be described mathematically. Although not all features show these patterns of Mendelian inheritance, his work suggested the utility of the application of statistics to the study of inheritance. Since that time many more complex forms of inheritance have been demonstrated.

The significance of Mendel's work was not understood until early in the twentieth century, after his death, when his research was re-discovered by other scientists working on similar problems.

Mendel did not understand the nature of inheritance. We now know that some heritable information is carried in DNA. (Retroviruses, including influenza, oncoviruses and HIV, and many plant viruses, carry information as RNA.) Manipulation of DNA can in turn alter the inheritance and features of various organisms.

LANDMARKS IN GENETICS

1859: Charles Darwin publishes The Origin of Species

1865: Gregor Mendel's paper, Experiments on Plant Hybridization

1869: Friedrich Miescher discovers a weak acid in the nuclei of white blood cells that today we call DNA (Hartl and Jones).

1880-1890: Walther Flemming, Eduard Strasburger, and Edouard van Beneden elucidate chromosome distribution during cell division

1903: Chromosomes are discovered to be hereditary units

1906: The term "genetics" is first introduced publicly by the British biologist William Bateson at the Third International Conference on Plant Hybridization in London, England.

1910: Thomas Hunt Morgan shows that genes reside on chromosomes, and discovered linked genes on chromosomes that do not follow Mendel's law of independent allele segregation

1913: Alfred Sturtevant makes the first genetic map of a chromosome

1913: Gene maps show chromosomes containing linear arranged genes

1918: Ronald Fisher publishes On the correlation between relatives on the supposition of Mendelian inheritance-the modern synthesis starts.

1927: Physical changes in genes are called mutations

1928: Frederick Griffith discovers a hereditary molecule that is transmissible between bacteria.

1931: Crossing over is the cause of recombination.

1941: Edward Lawrie Tatum and George Wells Beadle show that genes code for proteins; see the original central dogma of genetics

1944: Oswald Theodore Avery, Colin McLeod and Maclyn McCarty isolate DNA as the genetic material (at that time called transforming principle)

1950: Erwin Chargaff shows that the four nucleotides are not present in nucleic acids in stable proportions, but that some general rules appear to hold (e.g., the nucleotide bases Adenine-Thymine and Cytosine-Guanine always remain in equal proportions). Barbara McClintock discovers transposons in maize

1952: The Hershey-Chase experiment proves the genetic information of phages (and all other organisms) to be DNA

1953: DNA structure is resolved to be a double helix by James D. Watson and Francis Crick, with the help of Rosalind Franklin

1956: Jo Hin Tjio and Albert Levan established the correct chromosome number in humans to be 46

1958: The Meselson-Stahl experiment demonstrates that DNA is semiconservatively replicated

1961: The genetic code is arranged in triplets

1964: Howard Temin showed using RNA viruses that Watson's central dogma is not always true

1970: Restriction enzymes were discovered in studies of a bacterium, Haemophilus influenzae, enabling scientists to cut and paste DNA

1972: Walter Fiers and his team at the Laboratory of Molecular Biology of the University of Ghent (Ghent, Belgium) were the first to determine the sequence of a gene: the gene for Bacteriophage MS2 coat protein.

1976: Walter Fiers and his team determine the complete nucleotide-sequence of Bacteriophage MS2-RNA

1977: DNA is sequenced for the first time by Fred Sanger, Walter Gilbert, and Allan Maxam working independently. Sanger's lab complete the entire genome of sequence of Bacteriophage F-X174.

1983: Kary Banks Mullis discovers the polymerase chain reaction enabling the easy amplification of DNA

1985: Alec Jeffreys discovers genetic finger printing.

1989: The first human gene is sequenced by Francis Collins and Lap-Chee Tsui. It encodes the CFTR protein. Defects in this gene cause cystic fibrosis

1995: The genome of Haemophilus influenzae is the first genome of a free living organism to be sequenced.

1996: Saccharomyces cerevisiae is the first eukaryote genome sequence to be released

1998: The first genome sequence for a multicellular eukaryote, C. elegans is released.

2001: First draft sequences of the human genome are released simultaneously by the Human Genome Project and Celera Genomics.

2003: (14 April) Successful completion of Human Genome Project with 98% of the genome sequenced to a 99.99% accuracy

2006: Marcus Pembrey and Olov Bygren publish Sex-specific, male-line transgenerational responses in humans, a proof of epigenetics.

AREAS OF GENETICS

Classical Genetics

Classical genetics consists of the techniques and methodologies of genetics that predate the advent of molecular biology. After the discovery of the genetic code and such tools of cloning as restriction enzymes, the avenues of investigation open to geneticists were greatly broadened. Some classical genetic ideas have been supplanted with the mechanistic understanding brought by molecular discoveries, but many remain intact and in use, such as Mendel's laws and Muller's morphs. Patterns of inheritance still remain a useful tool for the study of genetic diseases.

Behavioral Genetics

Behavioral genetics studies the influence of varying genetics on animal behavior. Behavioral genetics studies the effects of human disorders as well as its causes. Behavioral genetics has yielded some very interesting questions about the evolution of various behaviors, and even some fundamental principles of evolution in general. For example, guppies and meerkats seem to be genetically driven to post a lookout to watch for predators. This lookout stands a significantly slimmer chance of survival than the others, so because of the mechanism of natural selection, it would seem that this trait would be lost after a few generations. However, the gene has remained, leading evolutionary philosopher/scientists such as Richard Dawkins and W. D. Hamilton to propose explanations, including the theories of kin selection and reciprocal altruism. The interactions and behaviors of gregarious creatures is partially genetic in cause and must therefore be approached by evolutionary theory.

Clinical Genetics

Physicians who are trained as Geneticists diagnose, treat, and counsel patients with genetic disorders or syndromes. These doctors are typically trained in a genetics residency and/or fellowship. Clinical genetics is also the study of genetic causes of clinical diseases.

Molecular Genetics

Molecular genetics builds upon the foundation of classical genetics but focuses on the structure and function of genes at a molecular level. Molecular genetics employs the methods of both classical genetics (such as hybridization) and molecular biology. It is so-called to differentiate it from other sub fields of genetics such as ecological genetics and population genetics. An important area within molecular genetics is the use of molecular information to determine the patterns of descent, and therefore the correct scientific classification of organisms: this is called molecular systematics. The study of inherited features not strictly associated with changes in the DNA sequence is called epigenetics.

Some take the view that life can be defined, in molecular terms, as the set of strategies which RNA polynucleotides have used and continue to use to perpetuate themselves. This definition grows out of work on the origin of life, specifically the RNA world hypothesis.

Population, Quantitative and Ecological Genetics

Population, quantitative and ecological genetics are all very closely related subfields and also build upon classical genetics (supplemented with modern molecular genetics). They are chiefly distinguished by a common theme of studying populations of organisms drawn from nature but differ somewhat in the choice of which aspect of the organism on which they focus. The foundational discipline is population genetics which studies the distribution of and change in allele frequencies of genes under the influence of the four evolutionary forces: natural selection, genetic drift, mutation and migration. It is the theory that attempts to explain such phenomena as adaptation and speciation.

The related subfield of quantitative genetics, which builds on population genetics, aims to predict the response to selection given data on the phenotype and relationships of individuals. A more recent development of quantitative genetics is the analysis of quantitative trait loci. Traits that are under the influence of a large number of genes are known as quantitative traits, and their mapping to a location on the chromosome requires accurate phenotypic, pedigree and marker data from a large number of related individuals. Ecological genetics again builds upon the basic principles of population genetics but is more explicitly focused on ecological issues. While molecular genetics studies the structure and function of genes at a molecular level, ecological genetics focuses on wild populations of organisms, and attempts to collect data on the ecological aspects of individuals as well as molecular markers from those individuals.

Population genetics is closely linked with the methods of genetic epidemiology. One method to study gene-disease associations is using the principle of Mendelian randomization.

GENOMICS

A more recent development is the rise of genomics, which attempts the study of large-scale genetic patterns across the genome for (and in principle, all the DNA in) a given species. The field typically depends on the availability of whole genome sequences, computational tools and Sequence profiling tool using bioinformatics approaches for analysis of large sets of data.

CLOSELY-RELATED FIELDS

The science which grew out of the union of biochemistry and genetics is widely known as molecular biology. The term "genetics" is often widely conflated with the notion of genetic engineering, where the DNA of an organism is modified for some kind of practical end, but most research in genetics is aimed at understanding and explaining the effect of genes on phenotypes and in the role of genes in populations, rather than genetic engineering.

MATRIX BIOLOGY

Matrix biology is the study of the structure and function of the extracellular matrix (ECM) and its interactions with proximal cells. Matrix biologists usually do both basic research into structure and cellular interactions of the ECM and applied research into how these parameters are altered in disease or other pathological states.

Matrix biologists can thus specialise in a particular class of molecule:

- collagen
- proteoglycans
- elastin
- integrins
- fibrillins
- matrix metalloproteinases

Specialities can also be by the particular ECM under study:

- cartilage
- skin
- bone
- tumours

Many matrix biologists also work closely with medical specialists and specialise in a particular disease or pathology:

- rheumatology-studying the change in ECM in cartilage and synovium in joints from rheumatoid or osteoarthritis patients
- orthopedics-studying pathological and traumatically injured cartilage and tendon
- oncology-studying how tumours and other cancers grow new ECM, metastasise and invade ECM of other tissues
- orthodontics-studying how bone remodels around teeth
- dermatology-studying ECM involved in wound healing and disease states such as scleroderma

There are many matrix biology societies who hold meetings regularly all over the world. These meetings generate a forum for matrix biologists to discuss common interests and form collaborations:

- ISMB-International Society of Matrix Biologists
- ASMB-American Society for Matrix Biology
- MBSANZ-Matrix Biology Society of Australia and New Zealand
- BSMB-British Society for Matrix Biology

MICROBIOLOGY

Microbiology is the study of microorganisms, which are unicellular or cell-cluster microscopic organisms. This includes eukaryotes such as fungi and protists, and prokaryotes such as bacteria and certain algaes. Viruses, though not strictly classed as living organisms, are also studied. People that study the field of microbiology are known as microbiologists.

Although much is now known in the field of microbiology, advances are being made regularly. The most common estimates suggest that we have studied only about 1% of all of the microbes in any given environment.

Thus, despite the fact that over three hundred years have passed since the discovery of microbes, the field of microbiology is clearly in its infancy relative to other biological disciplines such as zoology, botany and entomology.

History

Bacteria were first observed by Anton van Leeuwenhoek in 1676 using a single-lens microscope of his own design. The name "bacterium" was introduced much later, by Ehrenberg in 1828, derived from the Greek word *Bact* meaning "small stick". While Antony van Leeuwenhoek is often cited as the first microbiologist, the first recorded microbiological observation, that of the fruiting bodies of molds, was made earlier in 1665 by Robert Hooke.

The field of bacteriology (later a subdiscipline of microbiology) is generally considered to have been founded by Ferdinand Cohn (1828-1898), a botanist whose studies on algae and photosynthetic bacteria led him to describe several bacteria including Bacillus and Beggiatoa. Ferdinand Cohn was also the first to formulate a scheme for the taxonomic classification of bacteria.

Louis Pasteur (1822-1895) and Robert Koch (1843-1910) were contemporaries of Cohn's and are often considered to be the founders of medical microbiology. Pasteur is most famous for his series of experiments designed to disprove the then widely held theory of spontaneous generation, thereby solidifying microbiology's identity as a biological science.

Pasteur also designed methods for food preservation (pasteurization) and vaccines against several diseases such as anthrax, fowl cholera and rabies.

Robert Koch is best known for his contributions to the germ theory of disease, proving that specific diseases were caused by specific pathogenic microorganisms. He developed a series of criteria that have become known as the Koch's postulates. Koch was one of the first scientists to focus on the isolation of bacteria in pure culture resulting in his description of several novel bacteria including Mycobacterium tuberculosis, the causative agent of tuberculosis.

While Louis Pasteur and Robert Koch are often considered the founders of microbiology, their work did not accurately reflect the true diversity of the microbial world because of their exclusive focus on microorganisms having medical relevance.

It was not until the work of Martinus Beijerinck (1851-1931) and Sergei Winogradsky (1856-1953), the founders of general microbiology (an older term encompassing aspects of microbial physiology, diversity and ecology), that the true breadth of microbiology was revealed.

Martinus Beijerinck made two major contributions to microbiology: the discovery of viruses and the development of enrichment culture techniques. While his work on the Tobacco Mosaic Virus established the basic principles of virology, it was his development of enrichment culturing that had the most immediate impact on microbiology by allowing for the cultivation of a wide range of microbes with wildly different physiologies.

Sergei Winogradsky was the first to develop the concept of chemolithotrophy and to thereby reveal the essential role played by microorganisms in geochemical processes. He was responsible for the first isolation and description of both nitrifying and nitrogen-fixing bacteria.

TYPES OF MICROBIOLOGY

The field of microbiology can be generally divided into several subdisciplines:

- *Microbial Physiology*: The study of how the microbial cell functions biochemically. Includes the study of microbial growth, microbial metabolism and microbial cell structure.
- *Microbial Genetics*: The study of how genes are organised and regulated in microbes in relation to their cellular functions. Closely related to the field of molecular biology.
- *Medical Microbiology*: The study of the role of microbes in human illness. Includes the study of microbial pathogenesis and epidemiology and is related to the study of disease pathology and immunology.
- *Veterinary Microbiology*: The study of the role in microbes in veterinary medicine.
- *Environmental Microbiology*: The study of the function and diversity of microbes in their natural environments. Includes the study of microbial ecology, microbially-mediated nutrient cycling, geomicrobiology, microbial diversity and bioremediation. Characterisation of key bacterial habitats such as the rhizosphere and phyllosphere.
- *Evolutionary Microbiology*: The study of the evolution of microbes. Includes the study of bacterial systematics and taxonomy.
- *Industrial Microbiology*: The exploitation of microbes for use in industrial processes. Examples include industrial fermentation and wastewater treatment. Closely linked to the biotechnology industry. This field also includes brewing, an important application of microbiology.

- *Aeromicrobiology*: The study of airborne microorganisms.
- *Food Microbiology*: The study of microorganisms causing food spoilage.
- *Pharmaceutical Microbiology*: The study of microorganisms causing pharmaceutical contamination and spoillage.

BENEFITS OF MICROBIOLOGY

While microbes are often viewed negatively due to their association with many human illnesses, microbes are also responsible for many beneficial processes such as industrial fermentation (e.g. the production of alcohol and dairy products), antibiotic production and as vehicles for cloning in higher organisms such as plants.

Scientists have also exploited their knowledge of microbes to produce biotechnologically important enzymes such as Taq polymerase, reporter genes for use in other genetic systems and novel molecular biology techniques such as the yeast two-hybrid system.

MOLECULAR BIOLOGY

Molecular biology is the study of biology at a molecular level. The field overlaps with other areas of biology and chemistry, particularly genetics and biochemistry. Molecular biology chiefly concerns itself with understanding the interactions between the various systems of a cell, including the interrelationship of DNA, RNA and protein synthesis and learning how these interactions are regulated. Writing in Nature, William Astbury described molecular biology as:

> *"... not so much a technique as an approach, an approach from the viewpoint of the so-called basic sciences with the leading idea of searching below the large-scale manifestations of classical biology for the corresponding molecular plan. It is concerned particularly with the forms of biological molecules and..... is predominantly three-dimensional and structural-which does not mean, however, that it is merely a refinement of morphology- it must at the same time inquire into genesis and function."*

Relationship to other "molecular-scale" biological sciences.

Researchers in molecular biology use specific techniques native to molecular biology , but increasingly combine these with techniques and ideas from genetics, biochemistry and biophysics. There is not a hard-line between these disciplines as there once was. The following figure is a schematic that depicts one possible view of the relationship between the fields:

- Biochemistry is the study of the chemical substances and vital processes occurring in living organisms.
- Genetics is the study of the effect of genetic differences on organisms. Often this can be inferred by the absence of a normal component (e.g. one gene). The study of "mutants"- organisms which lack one or more functional components with respect to the so-called "wild type" or normal phenotype. Genetic interactions such as epistasis can often confound simple interpretations of such "knock-out" studies.
- Molecular biology is the study of molecular underpinnings of the process of replication, transcription and translation of the genetic material. The central dogma of molecular biology where genetic material is transcribed into RNA and then translated into protein, despite being an oversimplified picture of molecular biology, still provides a good starting point for understanding the field. This picture, however, is undergoing revision in light of emerging novel roles for RNA.

Much of the work in molecular biology is quantitative, and recently much work has been done at the interface of molecular biology and computer science in bioinformatics and computational biology. As of the early 2000s, the study of gene structure and function, molecular genetics, has been amongst the most prominent sub-field of molecular biology.

Increasingly many other fields of biology focus on molecules, either directly studying their interactions in their own right such as in cell biology and developmental biology, or indirectly, where the techniques of molecular biology are used to infer historical attributes of populations or species, as in fields in evolutionary biology such as population genetics and phylogenetics. There is also a long tradition of studying biomolecules "from the ground up" in biophysics.

TECHNIQUES OF MOLECULAR BIOLOGY

Since the late 1950s and early 1960s, molecular biologists have learned to characterize, isolate, and manipulate the molecular components of cells and organisms. These components include DNA, the repository of genetic information; RNA, a close relative of DNA whose functions range from serving as a temporary working copy of DNA to actual structural and enzymatic functions as well as a functional and structural part of the translational apparatus; and proteins, the major structural and enzymatic type of molecule in cells.

Expression Cloning

One of the most basic techniques of molecular biology to study protein function is expression cloning. In this technique, DNA coding for a protein of interest is cloned (using PCR and/or restriction enzymes) into a plasmid (known as an expression vector).

This plasmid may have special promoter elements to drive production of the protein of interest, and may also have antibiotic resistance markers to help follow the plasmid. This plasmid can be inserted into either bacterial or animal cells. Introducing DNA into bacterial cells is called transformation, and can be completed with several methods, including electroporation, microinjection, passive uptake and conjugation. Introducing DNA into eukaryotic cells, such as animal cells, is called transfection.

Several different transfection techniques are available, including calcium phosphate transfection, liposome transfection, and proprietary transfection reagents such as Fugene. DNA can also be introduced into cells using viruses or pathenogenic bacteria as carriers. In such cases, the technique is called viral/bacterial transduction, and the cells are said to be transduced.

In either case, DNA coding for a protein of interest is now inside a cell, and the protein can now be expressed. A variety of systems, such as inducible promoters and specific cell-signaling factors, are available to help express the protein of interest at high levels. Large quantities of a protein can then be extracted from the bacterial or eukaryotic cell.

The protein can be tested for enzymatic activity under a variety of situations, the protein may be crystallized so its tertiary structure can be studied, or, in the pharmaceutical industry, the activity of new drugs against the protein can be studied.

POLYMERASE CHAIN REACTION (PCR)

The polymerase chain reaction is an extremely versatile technique for copying DNA. In brief, PCR allows a single DNA sequence to be copied (millions of times), or altered in predetermined ways. For example, PCR can be used to introduce restriction enzyme sites, or to mutate (change) particular bases of DNA. PCR can also be used to determine whether a particular DNA fragment is found in a cDNA library. PCR has many variations, like reverse transcription PCR (RT-PCR) for amplification of RNA, and, more recently, real-time PCR (qPCR) which allow for quantitative measurement of DNA or RNA molecules.

GEL ELECTROPHORESIS

Gel electrophoresis is one of the principal tools of molecular biology. The basic principle is that DNA, RNA, and proteins can all be separated using an electric field. In agarose gel electrophoresis, DNA and RNA can be separated based on size by running the DNA through an agarose gel. Proteins can be separated based on size using an SDS-PAGE gel. Proteins can also be separated based on their electric charge, using what is known as an isoelectric gel.

Southern Blotting

Named after its inventor, biologist Edwin Southern, the Southern blot is method for probing for the presence of a specific

DNA sequence within a DNA sample. DNA samples before or after restriction enzyme digestion are separated by gel electrophoresis and then transferred to a membrane by blotting via capilliary action.

The membrane can then be probed using a DNA probe labeled using a complement of the sequence of interest. Most original protocols used radioactive labels, however now non-radioactive alternatives are available. Southern blotting is less commonly used in laboratory science due to the capacity of using PCR to detect specific DNA sequences from DNA samples. However, these blots are still used for some applications, such as measuring transgene copy number in transgenic mice, or in the engwineering of gene knockout embryonic stem cell lines.

Northern Blotting

The Northern blot is used to study the expression patterns a specific type of RNA molecule as relative comparison among of a set of different samples of RNA. It is essentially a combination of denaturing RNA gel electrophoresis, and a blot. In this process RNA is separated based on size and is then transferred to a membrane that is then probed with a labeled complement of a sequence of interest.

The results may be visualized through a variety of ways depending on the label used, however, most result in the revelation of bands representing the sizes of the RNA detected in sample. The intensity of these bands is related to the amount of the target RNA in the samples analyzed. The procedure is commonly used to study when and how much gene expressing is occurring by measuring how much of that RNA is present in different samples. It is one of the most basic tools for determining at what time certain genes are expressed in living tissues.

Western Blotting

Antibodies to most proteins can be created by injecting small amounts of the protein into an animal such as a mouse, rabbit, sheep, or donkey (polyclonal antibodies) or produced in cell culture (monoclonal antibodies). These antibodies can be used for a variety of analytical and preprative techniques. In Western blotting, proteins are first separated by size, in a thin gel sandwiched between two glass plates in a technique known as SDS-PAGE (for Sodium Dodecyl Sulphate Poly-Acrylamide Gel Electrophoresis).

The proteins in the gel are then transferred to a PVDF, nitrocellulose, nylon or other support membrane. This membrane can then be probed with solutions of antibodies. Antibodies that specifically bind to the protein of interest can then be visualized by a variety of techniques, including coloured products, chemiluminescence, or autoradiography.

Analogous methods to western blotting can also be used to directly stain specific proteins in cells and tissue sections. However, these immunostaining methods are typically more associated with cell biology than molecular biology. The terms "Western" and "Northern" are jokes: The first blots were with DNA, and since they were done by Ed Southern, they came to be known as Southerns.

I don't think Patricia Thomas, inventor of the RNA blot, which became known as a "Northern", actually used the term. . To carry the joke further, one can find reference in the literature to "southwesterns" (Protein-DNA interactions) and "farwesterns" (Protein-Protein interactions).

Arrays

A DNA array is a collection of DNA spots attached to solid support that is used to view the expression profile of many genes simultaneously. The arrays can contains anywhere from hundreds of relatively large spots (macroarrays) to tens of thousands of microscopic spots (microarrays). Arrays can also be tagged with antibodies to test a serum. There exists sandwich techniques where a test solution is exposed to the antibody in the array, and is then exposed to another set of antibodies which theoretically binds to the antibody bound to the antibody on the array (hence the sandwich).

Abandoned Technology

As new procedures and technology become available, the older technology is rapidly abandoned. A good example is methods for determining the size of DNA molecules. Prior to gel electrophoresis, with agarose and polyacrylamide, DNA was sized with rate sedimentation in sucrose gradients, a slow and labor intensive technology requiring expensive instrumentation; prior to sucrose gradients, viscometry was used. Aside from their historical interest, it is worth knowing about older technology as it may be useful to solve a particular problem.

History

Molecular biology was established in the 1930s, the term was first coined by Warren Weaver in 1938 however. Warren was director of Natural Sciences for the Rockefeller Foundation at the time and believed that biology was about to undergo a period of significant change given recent advances in fields such as X-ray crystallography. He therefore channeled significant amounts of (Rockefeller Institute) money into biological fields.

3

A BRANCH OF BIOLOGY

Botany is the scientific study of plantlife. As a branch of biology, it is also sometimes referred to as plant science(s) or plant biology. Botany covers a wide range of scientific disciplines that study the structure, growth, reproduction, metabolism, development and diseases of plants. The study of plants and botany began with tribal lore, used to identify edible, medicinal and poisonous plants, making botany one of the oldest sciences. From this the scope of botany has increased to include the study of over 550,000 kinds or species of living organisms.

SCOPE AND IMPORTANCE

As with other life forms in biology, plant life can be studied from different perspectives, from the molecular, genetic and biochemical level through organelles, cells, tissues, organs, individuals, plant populations, and communities of plants. At each of these levels a botanist might be concerned with the classification (taxonomy), structure (anatomy), or function (physiology) of plant life.

Historically, botany covers all organisms that were not considered to be animals. Some of these "plant-like" organisms include fungi (studied in mycology), bacteria and viruses (studied in microbiology), and algae (studied in phycology). Most algae, fungi, and microbes are no longer considered to be in the plant kingdom. However, attention is still given to them by botanists, and bacteria, fungi, and algae are usually covered in introductory botany courses.

The study of plants is important for a number of reasons. Plants are a fundamental part of life on earth. They generate the oxygen, food, fibres, fuel and medicine that allow higher life forms to exist. Plants also absorb carbon dioxide, a significant greenhouse gas, through photosynthesis. A good understanding of plants is crucial to the future of human societies as it allows us to:

- Feed the world
- Understand fundamental life processes
- Utilize medicine and materials
- Understand environmental changes

HUMAN NUTRITION

Virtually all food eaten comes from plants, either directly from staple foods and other fruit and vegetables, or indirectly through livestock or other animals, which rely on plants for their nutrition. In other words, plants are at the base of nearly all food chains, or what ecologists call the first trophic level. Understanding how plants produce the food we eat is therefore important to be able to feed the world and provide food security for future generations, for example through plant breeding. Not all plants are beneficial to humans, some weeds are a considerable problem in agriculture and botany provides some of the basic science in order to understand how to minimize their impact. However, other weeds are pioneer plants which start an abused environment back on the road to rehabilitation, underlining that the term weed is a very relative concept, and that broadly defined a weed is simply a plant which is too successful. Ethnobotany is the study of this and/or other relationships between plants and people.

BASIC LIFE PROCESSES

Plants are convenient organisms in which fundamental life processes (like cell division and protein synthesis for example) can be studied, without the ethical dilemmas of studying animals or humans. The genetic laws of inheritance were discovered in this way by Gregor Mendel, who was studying the way pea shape is inherited. What Mendel learned from studying plants has had far reaching benefits outside of botany. Additionally, Barbara McClintock discovered 'jumping genes' by studying maize. These are a few examples that demonstrate how botanical research has an ongoing relevance to the understanding of fundamental biological processes.

MEDICINE AND MATERIALS

Many medicinal and recreational drugs, like cannabis, caffeine, and nicotine come directly from the plant kingdom. Aspirin, which originally came from the bark of willow trees, is just one example. There may be many novel cures for diseases provided by plants, waiting to be discovered.

Popular stimulants like coffee, chocolate, tobacco, and tea also come from plants. Most alcoholic beverages come from fermenting plants such as barley malt and grapes. Plants also provide us with many natural materials, such as cotton, wood, paper, linen, vegetable oils, some types of rope, and rubber. The production of silk would not be possible without the cultivation of the mulberry plant. Sugarcane, rapeseed, soy and other plants with a highly fermentable sugar or oil content have recently been put to use as sources of biofuels, which are important alternatives to fossil fuels.

ENVIRONMENTAL CHANGES

Plants can also help us understand changes in on our environment in many ways.

- Understanding habitat destruction and species extinction is dependent on an accurate and complete catalog of plant systematics and taxonomy.
- Plant responses to ultraviolet radiation can help us monitor problems like the ozone depletion.
- Analyzing pollen deposited by plants thousands or millions of years ago can help scientists to reconstruct past climates and predict future ones, an essential part of climate change research.

- Recording and analyzing the timing of plant life cycles are important parts of phenology used in climate-change research.
- Lichens, which are sensitive to atmospheric conditions, have been extensively used as pollution indicators.

In many different ways, plants can act a little like the 'miners canary', an early warning system alerting us to important changes in our environment. In addition to these practical and scientific reasons, plants are extremely valuable as recreation for millions of people who enjoy gardening, horticultural and culinary uses of plants every day.

EVOLUTIONARY HISTORY

Early Botany (before 1945)

Among the earliest of botanical works, written around 300 B.C., are two large treatises by Theophrastus: On the History of Plants (Historia Plantarum) and On the Causes of Plants. Together these books constitute the most important contribution to botanical science during antiquity and on into the Middle Ages. The Roman medical writer Dioscorides provides important evidence on Greek and Roman knowledge of medicinal plants.

In 1665, using an early microscope, Robert Hooke discovered cells in cork, a short time later in living plant tissue. The German Leonhart Fuchs, the Swiss Conrad von Gesner, and the British authors Nicholas Culpeper and John Gerard published herbals that gave information on the medicinal uses of plants.

Modern Botany (since 1945)

A considerable amount of new knowledge today is being generated from studying model plants like Arabidopsis thaliana. This weedy species in the mustard family was one of the first plants to have its genome sequenced. The sequencing of the rice (Oryza sativa) genome and a large international research community have made rice the de facto cereal/grass/monocot model. Another grass species, Brachypodium distachyon is also emerging as an experimental model for understanding the genetic, cellular and molecular biology of temperate grasses. Other commercially important staple foods like wheat, maize, barley, rye, millet and soybean are also having their genomes sequenced. Some of these are challenging to sequence because they have more than two haploid (n) sets of chromosomes, a condition known as polyploidy, common in the plant kingdom. The "Green Yeast" Chlamydomonas reinhardtii (a single-celled, green alga) is another plant model organism that has been extensively studied and provided important insights into cell biology.

In 1998 the Angiosperm Phylogeny Group published a phylogeny of flowering plants based on an analysis of DNA sequences from most families of flowering plants. As a result of this work, major questions such which families represent the earliest branches in the genealogy of angiosperms are now understood. Investigating how plant species are related to each other allows botanists to better understand the process of evolution in plants.

STRUCTURAL BIOLOGY

Structural biology is a branch of molecular biology concerned with the study of the architecture and shape of biological macromolecules-proteins and nucleic acids in particular-and what causes them to have the structures they have. This subject is of great interest to biologists, because macromolecules carry out most of the functions of a cell, and because typically it only is by coiling into a specific three-dimensional shape that they are able to perform their functions. This shape, which is called the "tertiary structure" of a molecule, depends in a complicated way on the molecule's basic composition, or "primary structure."

Biomolecules are too small with the most advanced light microscopes. The methods that structural biologists use to determine their structures generally involve measurements on vast numbers of identical molecules at the same time. These methods include crystallography, NMR, ultra fast laser spectroscopy, electron microscopy, electron cryomicroscopy (cryo-EM), and circular dichroism. Most often researchers use them to study the static "native states" of macromolecules. But variations on these methods are also used to watch nascent or denatured molecules assume or reassume their native states.

A third approach that structural biologists take to understanding structure is bioinformatics to look for patterns among the diverse sequences that give rise to particular shapes. Researchers often can deduce aspects of the structure of integral membrane proteins based on the membrane topology predicted by hydrophobicity analysis.

In the past few years it has become possible for highly accurate physical molecular models to complement the in silico study of biological structures. Rapid prototyping technologies such as those used by 3D Molecular Design, or the creation of molecular models in glass by Luminorum Ltd, are notable examples of recent advances in this field.

TAXONOMY

Taxonomy is the practice and science of classification. The word comes from the Greek, taxis, 'order' nomos, 'law' or 'science'. Taxonomies, which are composed of taxonomic units known as taxa (singular taxon), are frequently hierarchical in structure, commonly displaying parent-child relationships.

Applications

Originally the term taxonomy only referred to the science of classifying living organisms (now known as alpha taxonomy); however, the term is now applied in a wider, more general sense and now may refer to a classification of things, as well as to the principles underlying such a classification.

Almost anything, animate objects, inanimate objects, places, concepts, and events, may be classified according to some taxonomic scheme.

The term taxonomy may also apply to relationship schemes other than parent-child hierarchies, such as network structures. Other taxonomies may include single children with multi-parents, for example, "Car" might appear with both parents "Vehicle" and

"Steel Mechanisms"; to some however, this merely means that 'car' is part of several different taxonomies.

A taxonomy might also be a simple organization of objects into groups, or even an alphabetical list. In current usage within "Knowledge Management", taxonomies are seen as slightly less broad than ontologies.

Mathematically, a hierarchical taxonomy is a tree structure of classifications for a given set of objects. It is also named Containment hierarchy. At the top of this structure is a single classification, the root node, that applies to all objects. Nodes below this root are more specific classifications that apply to subsets of the total set of classified objects. So for instance, in common schemes of scientific classification of organisms, the root is called "Organism" followed by nodes for the ranks: Domain, Kingdom, Phylum, Class, etc.

Taxonomy and Mental Classification

Some have argued that the human mind naturally organizes its knowledge of the world into such systems. This view is often based on the epistemology of Immanuel Kant. Anthropologists have observed that taxonomies are generally embedded in local cultural and social systems, and serve various social functions. Perhaps the most well-known and influential study of folk taxonomies is Émile Durkheim's The Elementary Forms of Religious Life.

Various Taxonomies

In alpha taxonomy, as the scientific classification of organisms, the system includes the root called "Organism" (as this applies to all living things, it is implied rather than stated explicitly), followed by the ranks: Kingdom, Phylum (plural: phyla), Class, Order, Family, Genus, Species, and Subspecies, with over 40 various other ranks sometimes inserted, such as subphylum, superorder, subfamily, subtribe, etc., to handle complex groups such as insects (more at: scientific classification or Linnaean taxonomy). In phylogenetic taxonomy (or cladistic taxonomy), organisms can be classified by clades, which are based on evolutionary grouping by ancestoral traits. By using clades as the criteria for separation, cladistic taxonomy, using cladograms, can categorize taxa into unranked groups.

In numerical taxonomy or taximetrics, the field of solving or best-fitting of numerical equations that characterize all measurable quantities of a set of objects is called cluster analysis.

Non-scientific Taxonomy

Other taxonomies, such as those analyzed by Durkheim and Lévi-Strauss, are sometimes called folk taxonomies to distinguish them from scientific taxonomies that claim to be disembedded from social relations and thus objective and universal. Baraminology, a creationist alternative to cladistics, is a system that emphasizes reproductive isolation to delineate clear baramins (sometimes, "holobaramins)" of living things. The baramins are then grouped into levels of polybaramins based on similarity and degree of common design. These polybaramins can be identified as families, orders, divisions, and so on. Baraminologists often regard mammals as a separate class from other vertebrates.

The neologism folksonomy should not be confused with "folk taxonomy" (though it is obviously a contraction of the two words). Those who support scientific taxonomies have recently criticized folksonomies by dubbing them "fauxonomies" (French word "faux" means "false"). The phrase "enterprise taxonomy" is used in business to describe a very limited form of taxonomy used only within one organization. An example would be a certain method of classifying trees as "Type A", "Type B" and "Type C" used only by a certain lumber company for categorising log shipments.

Physiology

Wöhler showed In 1828 that organic molecules, such as urea, can be created by synthetic means that do not involve life, and thus provided a powerful argument against vitalism. The first enzyme, diastase, was described in 1833, and scientists began to explore connections between chemistry and biology. Advances in microscopy also had a profound impact on biological thinking: in 1839, Schleiden and Schwann proposed the cell theory-that the basic unit of organisms is the cell and all cells come from preexisting cells. The cytologist Walther Flemming in 1882 was the first to demonstrate that the discrete stages of mitosis were not an artifact of staining, but occurred in living cells, and moreover, that chromosomes doubled in number just before the cell divided and a daughter cell was produced. In 1887 August Weismann proposed that the chromosome number must then be halved in the case of the sexual cells, the gametes. This was shortly proved to be the case and the process of meiosis began to be understood.

By the mid 1850s the miasma theory of disease was largely superseded by the germ theory of disease, creating extensive interest in microorganisms and their interactions with other forms of life. By the 1880s, bacteriology was becoming a coherent discipline, especially through the work of Robert Koch, who introduced methods for growing pure cultures on agar gels containing specific nutrients in Petri dishes. The long-held idea that living organisms could easily originate from nonliving matter (spontaneous generation) was attacked in a series of experiments carried out by Louis Pasteur, while debates over vitalism vs. mechanism continued apace.

Natural History and Natural Philosophy

Widespread travel by naturalists in the early-to mid-nineteenth century resulted in a wealth of new information about the diversity and distribution of living organisms. Of particular importance was the work of Alexander von Humboldt, which analyzed the relationship between organisms and their environment (i.e., the domain of natural history) using the quantitative approaches of natural philosophy (i.e., physics and chemistry). Humboldt's work laid the foundations of biogeography and inspired several generations of scientists.

Geology and Paleontology

The emerging discipline of geology also brought natural history and natural philosophy closer together; the establishment of the stratigraphic column linked the spacial distribution of organisms to the temporal distribution, a key precursors to concepts of evolution. Georges Cuvier and others made great strides in comparative anatomy and paleontology in the late 1790s and early 1800s.

In a series of lectures and papers that made detailed comparisons between living mammals and fossil remains Cuvier was able to establish that the fossils were remains of species that had become extinct-rather than being remains of species still alive elsewhere in the world, as had been widely believed. Fossils discovered and described by Gideon Mantell, William Buckland, Mary Anning, and Richard Owen among others helped establish that there had been an 'age of reptiles' that had preceded even the prehistoric mammals. These discoveries captured the public imagination and focused attention on the history of life on earth. Most of these geologists held to catastrophism, but Charles Lyell's influential Principles of Geology (1830) introduced uniformitarianism, a theory that explained the geological past and present on equal terms.

Evolution and Biogeography

The first to propose an evolutionary theory was Jean-Baptiste Lamarck; based on the inheritance of acquired characteristics (an inheritance mechanism that was widely accepted until the 20th century), it described a chain of development stretching from the lowliest microbe to humans.

The British naturalist Charles Darwin, combining the biogeographical approach of Humboldt, the uniformitarian geology of Lyell, Thomas Malthus's writings on population growth, and his own morphological expertise, created a more successful evolutionary theory based on natural selection. Similar evidence lead Alfred Wallace to independently reach the same conclusions. Though natural selection would not be accepted as the primary mechanism of evolution until well into the 20th century, most scientists were convinced of evolution and common descent by the end of the 19th century.

Alfred Wallace, building on earlier work by Humbolt and Darwin, made major contributions to biogeography by focusing on the distribution of closely allied species with particular attention to the effects of geographical barriers during his research in the Amazon basin and the Malay archipelago. He discovered the Wallace line dividing the fauna of the Malay archipelago between a zone allied with Asia and a zone allied with Australia. The orinthologist Philip Sclater, drawing on the work of Wallace and others, proposed a system of 6 major geographical regions to describe the distribution of bird species in the world. Wallace and others would in turn extend Sclater's system from birds to animals of all kinds.

Heredity and Development

The study of scientific study of heredity grew rapidly in the wake of Darwin's On the Origin of Species (1859) with the work of Francis Galton and the biometricians. The origin of genetics is usually traced to the 1866 work of the Austrian monk Gregor Mendel, who would later be credited with the laws of inheritance. However, his work was not recognized as significant until 35 years afterward. In the meantime, a variety of theories of inheritance (based on pangenesis, orthogenesis, or other mechanisms) were debated and investigated vigorously. Embryology and ecology also became central biological fields, especially as linked to evolution and popularized in the work of Ernst Haeckel.

Twentieth Century Biological Sciences

At the beginning of the 20th century, biological research was largely a professional endeavour. However, most work was still in the natural history mode, which emphasized morhphological and phylogenetic analysis over experiment-based causal explanations. However, anti-vitalist experimental physiologists and embryologists, especially in Europe, were increasingly influential. The tremendous success of experimental approaches to development, heredity, and metabolism in the 1900s and 1910s demonstrated the power of experimentation in biology. In the following decades, experimental work replaced natural history as the dominant mode of research.

Classical Genetics and Evolutionary Synthesis

1900 marked the so-called rediscovery of Mendel: Hugo de Vries, Carl Correns, and Erich von Tschermak independently arrived at Mendel's laws (which were not actually present in Mendel's work). Soon after, cytologists proposed that chromosomes were the hereditary material. Between 1910 and 1915, Thomas Hunt Morgan and his fly lab forged these two ideas-both controversial-into the "Mendelian-chromosome theory" of heredity. They quantified the phenomenon of genetic linkage and postulated that genes reside on chromosomes like beads on string; they hypothesized crossing over to explain linkage and constructed genetic maps of the fruit fly Drosophila melanogaster, which became a widely used model organism.

Building on his work on heredity and hybridization, Hugo de Vries proposed a mutation theory of evolution, which was widely accepted in the early 20th century. Lamarckism also had many adherents. Darwinism was seen as incompatible with the continuously variable traits studied by biometricians, which seemed only partially heritable. In the 1920s and 1930s-following the acceptance of the Mendelian-chromosome theory-a unification of the idea of evolution by natural selection with Mendelian genetics produced the modern synthesis; inheritance of acquired characters was rejected, while mutationism gave way as genetic theories matured. These ideas continued to be developed in the discipline of population genetics and in the second half of the century began to be applied in the new discipline of the genetics of behavior, sociobiology, and, especially in humans, evolutionary psychology.

Ecology, Ethology, and Environmental Science

In the early 20th century, naturalists were faced with increasing pressure to professionalize and add rigor and preferably experimentation to their methods (as the newly prominent laboratory-based biological disciplines had done). Ecology had emerged as a combination of biogeography with the biogeochemical cycle concept pioneered by chemists, and field biologists developed quantitative methods such as the quadrat and adapted laboratory instruments and cameras for the field to further set their work apart from traditional natural history. Zoologists and botanists did what they could to mitigate the unpredictability of the living world; new institutions like the Carnegie Station for Experimental Evolution and the Marine Biological Laboratory provided more controlled environments for studying organisms through their entire life cycles.

The ecological succession concept, pioneered in the 1900s and 1910s by Henry Chandler Cowles and Frederic Clements, was important in early plant ecology. G. Evelyn Hutchinson's studies of the biogeography and biogeochemical structure of lakes and rivers (limnology) and Charles Elton's studies of animal food chains set the pace for the kinds of quantitative methods that spread to the developing ecological specialties. Ecology became an independent discipline in the 1940s and 1950s after Eugene P. Odum synthesized many of the concepts of ecosystem ecology, placing relationships between groups of organisms (especially material and energy relationships) at the center of the field.

Microbiology, Biochemistry, and Molecular Biology

By the end of the 19th century all of the major pathways of drug metabolism had been discovered, along with the outlines of protein and fatty acid metabolism and urea synthesis. In the early decades of the twentieth century the role of minor components of foods in human nutrition, the vitamins, began to be isolated and synthesized. In the 1920s and 1930s, biochemists-led by Carl and Cori-began to work out many of the central metabolic pathways of life: the citric acid cycle, glycogenesis and glycolysis, and the synthesis of steroids and porphyrins. Between the 1930s and 1950s Fritz Lipmann and others established the role of ATP as the universal carrier of energy in the cell, and mitochondria as the powerhouse of the cell. Such traditionally biochemical work continued to be very actively pursued throughout the 20th century and into the 21st.

Following the rise of classical genetics, many biologists-including a new wave of physical scientists in biology-pursued the question of the gene and its physical nature. The adoption of simpler model systems like the bread mold Neurospora crassa made it possible to connect genetics to biochemistry, most importantly with Beadle and Tatum's "one gene, one enzyme" hypothesis in 1941. Research on even simpler systems like tobacco mosaic virus and bacteriophage, aided by the new technologies of electron microscopy and ultracentrifugation, forced scientists to re-evaluate the literal meaning of life; virus heredity and reproducing nucleoprotein cell structures outside the nucleus ("plasmagenes") complicated the accepted Mendelian-chromosome theory.

Oswald Avery showed in 1943 that DNA was likely the genetic material of the chromosome, not its protein; the issue was settled decisively with the 1952 Hershey-Chase experiment. By 1953 James D. Watson and Francis Crick showed that the structure of DNA was a double helix and showed its probable connection to replication. It was clear to many biologists that DNA sequence must somehow determine amino acid sequence in proteins; physicist George Gamow proposed that a fixed genetic code connected proteins and DNA. Between 1953 and 1961, there were few known biological sequences-either DNA or protein-but an abundance of proposed code systems, a situation made even more complicated by expanding knowledge of the intermediate role of RNA. To actually decipher the code, it took an extensive series of experiments in biochemistry and bacterial genetics, between 1961 and 1966-most importantly the work of Nirenberg and Khorana.

This discovery, and others, especially the discovery of the first restriction enzyme in 1968 and PCR in 1983 gave rise to the vigorous science that we know today as molecular biology.

In 1965 it was shown that normal cells in culture divide only a fixed number of times (the Hayflick Limit) then aged and died. About the same time, stem cells were shown to be exceptions to this rule and began to be studied in earnest. Toward the end of the century, totipotent stem cells came to be recognized as crucial for the understanding of developmental biology and raised hopes for new medical applications. In 1983 the unity of much of the morphogenesis of organisms from fertilized egg to adult began to be unraveled by the discovery of the homeobox genes, first in fruit flies, then in other insects and animals, including humans.

Starting in 1990 an important mechanism of gene regulation, RNA interference began to be understood and became an important laboratory technique to knockdown genes in order to determine their function in model organisms. In 2006 a large international consortium began a collaborative effort to enable researchers to conveniently obtain mice that have any one of its approximately 20000 genes knocked out. This US$100 million effort is the largest biological research endeavor since the human genome project.

The study of organisms, their reproduction, and the functions of their organs had come to be the study of molecules. Reductionism was triumphant. Even the methods of scientific classification of organisms, especially cladistics, began in the last quarter of the century to use RNA and DNA sequences as characters. By the mid 1980s even the overall division of the tree of life into three domains (as opposed to the classical two), the Archaea, the Bacteria, and the Eukarya, based on Woese's pioneering work on 16S rRNA sequencing, became generally accepted in the scientific community.

BIOTECHNOLOGY

While cloning in plants was known for millennia it was only in 1951 that the first animal, the tadpole, was cloned by nuclear transfer. Dolly, the sheep was cloned by transfer of a mature somatic cell nucleus into an enucleated oocyte in 1997. Within a few years, several other animals, including dogs, cats, horses and cattle were cloned by similar methods.

The largest, most costly single biological study ever undertaken, the Human genome project began in 1988 under the leadership of James Watson, and a first draft of the human DNA sequence was announced in 2000. By 2003 99% of the genome had been sequenced to an accuracy of one part in ten thousand. The HapMap project to determine patterns of differences in the human genome began in 2002 and by 2005 completed its first phase work by of discovering on the order of one million SNPs in 270 people sampled from four distinct populations of people: Han Chinese, Japanese, Yoruba Nigerians, and Northern Europeans. The advent of whole-genome sequencing and surveys of their variation in different populations (races), together with new statistical methods, permitted researchers by 2006 to systematically identify candidate loci for recent natural selection during evolution in humans. Some of these genes were also shown to be ancestry-informative markers which came to be used in genealogical studies and to understand ancient human migrations.

The first genome of a plant model organism, Arabidopsis thaliana was sequenced in 2000. Dozens of bacteria, the mouse, the nematode Caenorhabditis elegans and other model organisms were sequenced and their genes mapped, often by large international collaborations. The first decade of the twenty-first century saw the rise of proteomics, computational biology and bioinformatics, with an emphasis on huge databases of experimentally derived data, all connected by the Internet and available to researchers everywhere, which has fundamentally changed the structure of the science of biology itself.

Household and Gender in a Life-course Perspective

One of the major developments in social science research over the past two decades has been the shift away from static approaches towards perspectives and methods that can shed light on the true flux of life. Of course, social change has long been a principal focus of social science. What is new is the increased attention to the interplay between the larger, societal forces of change and the actual experiences of people. Social scientists are moving away from simple models of "before" and "after" from traditional to modern, from pre-industrial to industrial, from rural to urban towards much more subtle formulations of how people's lifes are lived and how the course of their lives may vary as a result of macro-level developments.

The life-course perspective which arose from a confluence of movements in sociology and psychology in the 1960s subsequently came to involve economists, anthropologists, historians, and others. It looks at the distinctive series of roles and experiences through which the individual passes as she or he ages from birth to death and inquires into the impact of various changes on these patterns. Moreover, it asks how people's reactions to change affect larger societal forces which in turn influence the course of social change. In short, the perspective calls attention not only to the ways in which people's lives are influenced by broad economic, political, social, and cultural developments, but also to how the collective impact of individuals' reactions to these changes affects the course of subsequent change at the macro-level.

The Household, Gender, and Age (HGA) project of the United Nations University was designed as an attempt to provide better information on just how the many macro-level changes taking place in the developing countries of the world are affecting individuals in their family environment. The emphasis was on the experience of women, whose lives have often been little understood by the technicians and politicians most closely involved in development policy decisions.

Given these goals, the selection of a life-course approach to provide a common methodological framework for the individual country research projects made much sense. The life-course approach provides a sophisticated framework for empirical research without forcing specific, substantive questions on researchers whose regional and national settings may be very different. Hence, the individual researchers were able to formulate their own specific questions according to their local situation while still sharing a common, broader perspective on the relationship of macro-level change to the domestic situation of women. The life-course approach also encouraged the use of a wide range of data collection techniques and analysis, ranging from traditional life-history and other qualitative approaches to some of the most advanced and newly developed methods of longitudinal data-gathering and statistical analysis.

By assuring a dual research focus-the macro-forces affecting women's lives and the changing ways in which those lives are actually being lived-the life-course perspective provides a firm basis for understanding the human consequences of development.

As the HGA researchers have so richly documented, many of these consequences are unintended, and some have been disastrous.

One of the major advantages of a life-course approach is the key it provides to understanding the various consequences of social change at different stages of a person's life. All too often development specialists and social scientists speak of the impact of macro-level changes as though everyone in a community equally were affected regardless of gender, age, or situation. However, the point of the life-course which the individual has reached when a change is introduced may have profound implications for the effect of that change on that person. This is obvious in the case of introducing a new school, for example, as this will presumably directly affect youngsters more than the others in a community. Virtually all types of externally induced changes are like this The result is that reference to positive versus negative changes are often misleading, since some changes may have positive implications for those at a certain point in their life-course and negative implications for those at other points.

A final reason for adopting a life-course perspective in the HGA project was that it provides valuable insights on domestic group processes, the "H" of the HGA. The household is not a monolithic unit, despite the fact that planners and social scientists often speak of it as such. Rather, the household should be understood as the continuously changing product of the interaction of the group of individuals of whom it is comprised. Interests vary: those of the male head of household may differ from his wife's; in a polygynous household, the interests of the various wives may differ from one another; the interests of the oldest child may differ from that of younger siblings. The life-course perspective brings this to light, showing the actual complexity of interacting human lives that lies behind the misleading unity of the household label.

The essence of a life-course approach lies in the continuous interplay between social change and the life-course of individuals. People begin their lives in one historical period distinguished by a characteristic set of cultural norms and perceptions and institutional arrangements, and, as they age, these larger forces change. Individuals begin to develop ideas of normative family behaviour and the proper trajectory of their lives at an early age, but when they come to face life later it may be in a society very different from the one from which these ideas and norms first sprung. For example, they develop their ideas of what the life of a grandparent is like and what the relations between grandparents and their children and grandchildren are like when they arc children, yet when they later become grandparents themselves they may be living in a very different society, where being a grandparent could have a very different significance (Riley, 1985).

A life-course approach implies the notion of a society that is structured by age. That is, roles in a society differ by age, so that one can imagine a society that is structured by age in much the same way as it is structured by class and gender. Each society has its own cultural timetable for the appropriate progress of the individual through the life-course. There are social pressures which encourage individuals to make transitions between roles (for example, from unmarried to married) at points that arc defined as culturally appropriate (Hagestad and Neugarten, 1985).

Scholars have long been interested in the life-cycle, the sequences of roles through which the individual passes from birth to death. However, there has been an unfortunate tendency to look at the range of current life-stages-infancy, childhood, adolescence, young adulthood, and so on-and assume that together they represent the stages of life through which any individual in the society passes. As a life-course perspective stresses, this would be true only in the case of a society that was entirely unchanging. Since no current societies arc known to fall into this category, this means that we cannot generalize from the synchronic description of individual life-stages to characterize the pattern of roles and experiences through which any individual (or any cohort) passes over the life-course. The life-course approach, then, puts a sense of change and history into life-cycle research. A central concept in life-course analysis is the cohort or group of individuals who enter a given system or a given status at a particular time. A marriage cohort, for example, refers to all those who got married in a certain period. A birth cohort, which is the grouping used in this study, consists of all those who were born in the same time-span. Since different cohorts encounter different segments of time as they proceed through their lives, each is affected differently by historical change. One of the major aims of life-course research is to determine how changes are reflected in the way different cohorts age, and how the characteristics of different cohorts (their size, for example) may bring about historical changes (Riley et al., 1972).

This does not mean, however, that cohorts should be seen as homogeneous units. Cohort analysis involves not only comparison between different cohorts, but also the examination of variation within each cohort. Since members of the same cohort have different characteristics, they may be affected differently by the same historical change. For example, the introduction of mandatory elementary education may have a greater effect on women farmers or small traders, who depend on the labour of their young children, than on elite women, who do not.

As this example indicates, changes affecting one cohort (in this case children at the time elementary education becomes compulsory) can have important implications for the life-course pattern of other cohorts (in this case, the mothers of these children). Similarly, wage-labour opportunities for young men may have a major impact on the status of the elderly, whose control over their juniors may be diminished. This linkage between cohorts is a major focus of lie-course research.

Life-course analysis introduces a historical perspective into the study of how people live from birth to death. When looking at the household-one of the primary links of social organization-the life-course approach forces us to see the real dynamism of people's lives and to avoid the reification of academic analytical categories. The constellation of people who compose any household is the result of several different factors, including cultural norms about what is an appropriate place to live, demographic variables such as the survival of parents, number of siblings, children, and grandchildren, and economic factors, like the advantages of housing large numbers of related adults together where the household is also the productive unit.

Any individual's co-residential experiences are the result of the interaction of these different forces, which themselves are influenced by larger sources of historical change. Cultural norms regarding polygyny, for example, may change over time; decreasing mortality and changes in fertility and migration affect the demographic component of household composition; and changes in the economy lead to different pressures on households, which has a bearing on the advantages or disadvantages of large, complex family households.

A well-entrenched sociological approach to the study of the domestic living unit focuses on the developmental cycle of households, often identified with the family cycle. A major problem with the family-cycle approach is that in concentrating on how household units change over time, it often loses sight of what happens to individuals. For example, a family-cycle model may focus on the cycle involving the death of the elder generation and the passing on of the household to the eldest son. But what of the daughters and the younger sons? And what of those who had only daughters and no sons, or those whose parents died when they were young children? In short, the family-cycle approach tends to yield a picture of uniformity that is rarely found in practice (Kertzer, 1986).

By focusing on the actual experiences of individuals from birth to death, the life-course approach allows the research to capture the full diversity of their co-residence in households. It provides a better position to examine the impact of various changes such as those introduced by development schemes-on people's domestic lives. This cannot be done by simply following household units: it is necessary to look at the history of every individual over time, examining the choices he or she made and the changing economic, demographic, environmental, and cultural forces that influenced the person's decisions and experiences. The life-course approach to co-residence is especially useful for studying in developing countries, where the Western notions commonly built into family-cycle models are least appropriate. This is most notable in the many cases where polygyny is a cultural norm, for family-cycle research is based almost entirely on the assumption of a simple marital dyed: a single husband with a single wife. Using a life-course approach to co-residence it is possible to follow the co-residential experiences of each individual in whatever kind of household she or he lives.

A life-course approach provides a broad framework for the study of how women's lives and family relations generally have changed in the third world. By focusing attention on the interrelationship between the individual and the ever-changing society, it avoids the pitfalls of examining only macro-level phenomena (such as economic development per se) on the one hand, or, on the other, the particular experiences of the individual unrelated to the broader societal context. Moreover, unlike a simple life-history approach, it provides a means for aggregating the experiences of individuals to test propositions about the nature of the change that is taking place.

While the life-course approach provided a common framework for HGA researchers working in different countries, it did not constrain them to limit their specific substantive focus. Rather, each team could identify developmental and other issues important in their particular societal context and use a life-course framework to design their study of them.

The issues differed widely from country to country, but what unites the various studies of the HGA project is, on the one hand, a common interest in how larger societal changes influence the lives of women in their domestic context, and, on the other, the use of a life-course perspective to provide a common means for determining the nature of these changes and their implications.

The studies are also characterized by the multiple levels of comparison that are built into a life-course approach. These involve comparison between cohorts and between communities. The latter is, of course, a standard method in social science. For example, in the original Chinese report there were two levels of geographically based comparison. The first involved comparison between an area (in Sichuan province) that is less industrially and commercially developed than another (in Jiangsu province). The second involved comparisons within each area between larger, more centrally located communities and those that are smaller and more remote.

The distinctive feature of the comparisons in the HGA studies, though, is the use of cohort comparisons in addition to those geographically based. In order to trace the path of changes in people's lives, each national study involved selecting specified younger and older cohorts and comparing their life-trajectories. Such comparisons show how the life-courses of successive cohorts have varied as they have experienced change at different stages.

The method of cohort comparison employed in the HGA studies can be illustrated by reference to figure 1, which provides a hypothetical example of a comparative study of three birth cohorts. The movement of historical time is represented from left to right, with the passage of years shown at the bottom of the figure. The oldest cohort is cohort 1, the members of which were born around 1920, and whose last members are expected to die around 1995, when they are 75. The middle cohort, cohort 2, represents those born 20 years later, around 1940, while the youngest cohort, cohort 3, was born around 1960. The details of the cohort comparison method are simplified here for ease of explanation. Actual studies involve more broadly defined cohorts, for example, those born over a five-or ten-year period. Moreover, the more recent cohorts might be expected to have a longer life-span than the older cohorts.

As the members of each cohort age, they pass from one stage of their life-course to a successive stage. For convenience, four general life-stages have been distinguished: child (0-16), young adult (17-39), mature adult (40-59) and old age (over 60). Of course, life-stages are themselves culturally defined and not biologically fixed; indeed, a common theme in life-course research is the question of how a people divide up the life-course (Keith, 1985; Kertzer and Keith, 1984). The four divisions found in figure

1 are therefore not to be interpreted as cross-culturally valid categories, but simply as a convenient device for explaining the cohort comparison method. At any given date, members of different cohorts are of different ages. Examples of such dates are by vertical lines in figure 1. For instance, by following the vertical line from 1970, it is clear that members of cohort I were mature adults, members of cohort 2 were young adults, and members of cohort 3 were children in that year A historical event affects members of the different cohorts at different stages of their lives and so potentially has a different effect on each of them.

For example, a general famine in 1970 might have led to malnutrition and increased disease and mortality for all cohorts. This would be a period effect, affecting all cohorts in the same way. However, such a historical event would also be likely to have different long-term implications for the various cohorts. For instance, malnutrition at an early age may lead to a higher incidence of certain disabilities later in life with the result that the old-age experience of cohort 3 after the year 2000 would be different from the old-age experiences of cohort 1 in the 1980s and 1990s. This would represent a cohort effect.

In life-course studies, the focus is on how cohorts differ from one another as a result of historical change. If there were no change, presumably the experience of cohort I would be identical to that of cohorts 2 and 3. Their childhood and young adult experiences would all be the same, despite the fact that they took place in different periods. It was just such an assumption of an unchanging society that allowed earlier anthropologists and others to write studies of the life-cycle in traditional societies divided into chapters on infancy,. childhood, marriage, and old age. However, it is clear that such an assumption can no longer apply to any society. It is not possible to generalize on various life-stages in such a simple way, since the same life-stage differs from one cohort to another.

Moving from this rather general theoretical level to considering how the HGA researchers could put this method into practice shows that what may seem to be a straightforward conceptual framework encounters complexities when it comes to the research itself. Consider, for example, a comparison between a group of women who are in their fifties and a group of women in their thirties. If the findings suggest that the older women were more oriented to family life than the younger women, who spent more time in labour outside the household, it might be tempting to conclude that this represents a life-course pattern and that young women go through a period of involvement in the outside labour force which they leave as they grow older. In fact, many earlier analysts made just such assumptions.

However, there are other possible interpretations of these findings which would acknowledge historical change in life-course patterns and hence different experiences for different cohorts. If, for example, a textile factory favouring young female workers had been established in the community in recent years, it could be that the pattern of extra-domestic labour for young women is new, and that the women in the older cohort never passed through such a stage. The life-course approach is sensitive to these important differences of interpretation and is designed to distinguish between such competing explanations of life-course patterns. Cohort comparison, then, is comparing segments of the life-course which are experienced by different cohorts in different historical periods. For example, returning to figure 1 to examine the course of women's lives as young adults would involve investigating the experiences of cohort 1 in the period 1937-1960 and comparing it with the experiences of cohort 2 in 1957-1980 and cohort 3 beginning in 1977. This would reveal the results of historical change in this phase of the life-course of all the women concerned.

It should now be evident that a lie-course approach requires longitudinal data, that is, information on individuals over time, beginning at their birth and ideally continuing on until they die. This contrasts with most social research, which has relied heavily on collecting cross-sectional data on a population at a certain point in time. This is the typical approach in most survey research. The need for longitudinal data in life-course research places heavy demands on the researcher. There are a variety of ways in which life-course data can be gathered. It can be collected through archive-based studies that use governmental, church, family, or other organizational records to reconstruct the lives of individuals. It can be done through prospective or panel studies that begin at a certain date and then continually return to the subject population to follow them as they age. Thirdly, there are retrospective studies that take place at a certain point in time but try to elicit information from individuals on their entire life-course from birth to the date of the interview. In the HGA, the last of these alternatives was adopted, since in many settings sufficient historical records for the first approach were not available, while the second approach would require a commitment to a multi-year (if not multi-decade) study which would take a long time to produce results and which, in any case, would require a massive budget.

The design adopted by the HGA resulted in the use of a common instrument in most of the component projects. This was a life-course matrix that was based on retrospective information on various dimensions of the lives of those interviewed, such as marriage, work, education, co-residence, and fertility. In a work history, for example, the person would be asked when she first went to work and what kind of job it was. She would then be asked the date or her age when she left that job and the reason she did so, followed by questions about her next job. Her work history, then, would consist of the dated series of jobs she held from the time she began work to the present.

The vertical dotted line in figure I represents a date (1985) for the collection of such retrospective life-course data. Note that the data collected for cohort 1 are much more complete than those gathered for cohort 2, which, in turn, are fuller than data gathered for the youngest cohort, cohort 3. In fact, the members of cohort 3 are only about 25 years old at the time of the interview, and so nothing can be known about their mature adult or older years, and it is impossible to compare all three cohorts for this portion of the life-course. It is possible to compare all three cohorts' childhood experiences, analysing historical differences and their causes, and it is also possible to compare the early years of young adulthood of all three cohorts. However, a study of mature adulthood is limited to a two-way comparison of cohorts 1 and 2.

One other methodological note to be kept in mind in conducting retrospective life-history research concerns the problem of selectivity bias. The older the members of the cohort being interviewed, the more likely they arc to be a highly selective subset of the original cohort. This is because survival to old age is not randomly distributed among members of a cohort, but is linked to a variety of factors in earlier life. For example, interviewing all living members of cohort 1 in 1985 means interviewing those who had survived to age 65. This could well be a minority of all those who had been original cohort members in 1920, since most of them might have died. Therefore any picture of the childhood experiences of cohort l based entirely on the retrospective reports of those members who were interviewed in 1985 is likely to reach erroneous conclusions if there is no allowance for selectivity bias.

To give an extreme example, imagine a society with no historical change, with each cohort following the same path and having the same characteristics over the life-course as every other cohort. Imagine, though, that half the population is poor and has high infant and childhood mortality rates. This half also has different kinds of early life-course experiences, for example more time in the labour force, than the richer half. Comparing the older cohort to the younger cohort on the basis of retrospective reports of what the youth of each was like would lead to the conclusion that the two cohorts had very different experiences, even though, in fact, there were no differences at all. Since few of the poorer people would have survived to be interviewed at 65, the picture of the childhood of the older cohort would be based largely on the experiences of the wealthy half of the population. There is no space here to go into the details of how such problems need to be handled. However, the problem of selectivity bias must be carefully dealt with in any analysis of retrospective data.

One of the valuable aspects of the life-course approach in the context of the HGA project is that it called for both qualitative and quantitative data. This, too, is reflected in the different individual country projects. While the life-history matrix described above calls for quantitative data, the interpretation of life-course experiences benefits greatly from supplementary qualitative information. Three kinds of approach-open-ended survey questions, participant observation, and life-histories are particularly useful here.

Open-ended survey questions complement close-ended questions on most surveys. By providing an opportunity for the person to respond in her own terms to a broad area of inquiry, such questions allow the researcher greater insight into how people perceive their own lives. They also suggest causal chains that the researcher might not have previously considered, some of which might subsequently be through more close-ended questioning of a broader sample.

Possibilities for participant observation varied from project to project, but it was possible to some degree in virtually all of them. Sharing the daily life experiences of the people in the study provides a richer understanding of their lives. Direct observations of behaviour lead to questions that might not otherwise arise, and observation also allows the researcher to distinguish between what people claim happens in certain situations and what in fact actually happens. More generally, sharing the daily lives of those concerned leads to greater acceptance by the local people and results in better-quality data, including those obtained through the administration of the more formal research instruments.

All of the HGA projects collected some intensive life-histories. In general, an attempt was made to collect life-histories from individuals representing different cohorts and different subgroups in each country study. Such information not only provides richer detail about the course of people's lives, but also allows for people's own interpretation of their lives and of the historical events that affected them. In the end, the best life-course analysis is often the one that can blend these various instruments to describe people's changing lives, to identify the factors that have led to the changes, and to discover the ways in which these changes are viewed by the people themselves.

Although the individual country studies vary in their emphasis on qualitative versus quantitative data, all of them did collect systematic quantitative data on life-course processes. While these data provide many advantages over simple survey-style cross-sectional data, their analysis does introduce new complexities associated with their longitudinal nature. Fortunately, the statistical analysis of longitudinal data has been one of the areas in the social sciences most actively developed in recent years, especially in relation to sociology and economics. Treatment of the technical issues involved in event-history analysis and other forms of statistical analysis for life-course data is beyond the scope of this chapter. However, the reader is referred to such works as Tuma and Hannan (1984), Allison (1984), and Kalbfleisch and Prentice (1980) for a presentation of many of the relevant statistical methods. Dennis P. Hogan, professor of sociology at Pennsylvania State University, served as principal consultant on analytical methods for the HGA project.

The use of quantitative life-course data also introduces complexities in data management. Event-history data, like that gathered for the HGA projects, entail collecting information on all relevant events like births, marriage and marriage dissolution, new jobs, and unemployment for each individual in the study throughout her life-course, that is, from birth to the time of the interview. This leads to a form of data very different from that found in typical cross-sectional studies. The typical survey, for example, consists of a set of questions, each of which has a list of possible responses. The resulting data set has a regular rectangular form: each variable occupies a certain column or columns, and each person contributes a single record to the database.

Event histories, on the other hand, do not have this simple structure. This is partly due to the fact that the lives of some people are more eventful than others. Person A may have had 20 different jobs while person B has never been in the workforce. At the same time, person A may never have been married, while person B has been married and divorced four times. To deal with an event-history database as if it were a cross-sectional database would require allowing, for each such variable as jobs and marriage, as many occurrences as are found in the life of the person with the most such events (the most jobs, the most marriages, and so on). The result would be an enormous database consisting very largely of blanks. If, for example, just one out of 1,000 respondents had had 40 different jobs, there would have to be space in each of the 1,000 records for information on 40 jobs, even though most people had no more than three.

The solution adopted by many of the HGA projects was a computer database management system devised by Nancy Karweit, a sociologist at the Centre for Social Organization of Schools at Johns Hopkins University in Baltimore. The nature of this system is described in Karweit and Kertzer (1986). It involves dividing the life-course data into different domains (marriage, fertility, employment, and schooling, for example) and providing a separate record for every single event occurring to a person in each domain. Hence a person who married three times would have three marriage records, while a person who never married would have none. More than a way of simply economizing space in the data file, this system, known as CASA, provides a means for retrieving data to capture life-course processes and to permit statistical analysis. Versions of CASA have been prepared to run on both mainframe and IBM-compatible microcomputers.

With the tremendous changes that have been taking place in the developing countries of the world, interest has inevitably gone beyond broader questions of economic development, demographic change, and political revolution to the question of what these macro-forces mean to the people, to individuals. In looking at how lives have been changed by these historical developments, it is necessary to move beyond conventional sociological models which seek to ferret out relationships between variables (such as occupation and fertility) in a system that is assumed to be unchanging. A life-course perspective concentrates on the link between the larger field of economic, political, demographic, and institutional change on the one hand and the lives of individuals on the other. It avoids analyses which ignore the actual life experience of individuals and also goes well beyond micro-level studies that fail to relate the lives of individuals to the changing constellation of macro-forces that mould individual lives and lead to different patterns of experience for successive cohorts.

Life-course study represents a blending of old and new methods, of quantitative and qualitative instruments. These projects focused on the lives of women, but the life-course approach meant that their lives could only be validly seen in terms of the broader social context-as represented in marriage and kinship relationships, for example-that included men as well as women. While focusing much on how larger social and economic factors have altered the lives of women in their household setting, the life-course approach also reveals how individual women have responded creatively to these external pressures in fashioning satisfactory lives for themselves and their families. These innovations, introduced from below, in turn influence the course that younger women will follow later, thus having an impact on the course of economic development and the path of social change.

4

THE SCIENTIFIC PROGRESS OF BIOLOGY

INTRODUCTION

The concept of "scientific progress", which dates from the sixteenth and seventeenth centuries, has long been linked with an ideal of free and open dissemination of scientific information. The discipline of sociology is much newer, but since the first half of the twentieth century sociologists of science have attemped to cast light on the connection between information flow and ongoing research and development.

Early sociologists of science placed great importance on the freedom of scientific communication and exchange. Legal protection of intellectual property rights was specifically condemned as a violation of scientific cultural norms which were supposed to align the self-interest of individual researchers with the common goal of scientific progress. Although most contemporary sociologists of science would not accept the existence of a set of functional scientific norms, up-to-date research in the sociology of science suggests that the exchange of information among scientists is an important aspect of scientific research which may be adversely affected by the expansion of intellectual property rights.

This section outlines theoretical developments in the sociology of science relating to the nature and significance of scientific exchange, beginning with the theories of Robert K. Merton and his colleagues and moving on to identify a useful and up-to-date sociological framework for the present study.

Traditional sociology of science. Merton's norms of science.

Why did early sociologists of science consider free scientific exchange to be critical for scientific progress?

The concept of "scientific progress" dates from the sixteenth and seventeenth centuries and has long been linked with an ideal of free and open dissemination of scientific information. During the 1940s, sociologists of science formalised this conceptual link by theorising that a norm of common ownership of research results--the norm of "communism", or "communalism"--functioned together with other scientific cultural norms to align the interests of individual scientists with the overarching institutional goal of scientific progress, defined as the extension of certified knowledge. According to sociologist Robert K. Merton and others of his school, these norms were not codified or necessarily explicit; rather, they operated as "prescriptions, proscriptions, preferences and permissions... legitimated in terms of institutional values... transmitted by precept and example and reinforced by sanctions". Their existence could, it was argued, be inferred from a moral consensus among scientists expressed "in use and wont, in countless writings on the scientific spirit and in moral indignation directed toward contravention of the ethos". Disagreement among scientists was acknowledged by sociologists, but regarded as deviant and generally attributed to insufficiently internalised norms.

The "norms of science", especially the norm of communism, reflected what sociologists regarded as the essentially cooperative and cumulative nature of scientific research. In order to collaborate and build on each other's work, scientists needed access to a common fund of knowledge. The norm of communism was supposed to encourage scientists to contribute to this common fund by communicating the results of their research to other scientists: the norm ensured that secrecy was condemned, while timely, open publication was rewarded. The norm was also supposed to preserve scientific knowledge within the public domain, where it could be freely used and extended. Merton wrote: "The substantive findings of science... constitute a common heritage in which the equity of the individual producer is severely limited. An eponymous law or theory does not enter into the exclusive possession of the discoverer and his heirs, nor do the mores bestow upon them special rights of use and disposition. Property rights in science are whittled down to a bare minimum by the rationale of the scientific ethic. The scientist's claim to "his" intellectual "property" is limited to that of recognition and esteem".

Like the ideal of open dissemination of scientific knowledge, the notion that science should be commonly owned and pursued for the public good has long been associated with the concept of scientific progress. In the 1930s and 1940s, leaders of the "radical science movement", from which the research agenda adopted by early sociologists of science evolved, set out to consider how science could best be organised for maximum social benefit. Thus, Merton intended his examination of the normative structure of science as an introduction to a larger problem: the comparative study of the structure of science under different political conditions. He argued that because the institution of science is only a part of the larger social structure, with which it is not always integrated, the scientific ethos can be subjected to serious strain when the larger culture opposes a scientific norm. In the case of the norm of communism, Merton saw such conflict arising out of the incompatibility of the scientific norm with the definition of technology as private property in a capitalist economy. He referred specifically to patents, with their exclusive rights of use (and, he remarked, often non-use), and to the suppression or withholding of knowledge (for example, through trade secrecy) as being in opposition to the rationale of scientific production and diffusion.

MORE RECENT APPROACHES

Current Interpretations of the "Norms of Science"

The first genetically engineered organisms were created in the United States in 1973 by academic scientists Herbert Boyer and Stanley Cohen. By this time, mainstream sociologists of science had largely abandoned models of scientific research based on Merton's theory of scientific norms.

The reason relates to the historical development of the discipline: early philosophers and sociologists of science regarded science as unique among intellectual activities, and in particular, they thought that science as a discipline was defined by a high level of agreement among scientists about assertions of fact; for this reason early sociologists of science, including Merton, were chiefly concerned with constructing models to explain the phenomenon of scientific consensus. However, during the 1960s and 1970s--influenced by developments in the history and philosophy of science--sociologists of science began to take a more cynical view of competition and collaboration within scientific communities.

Questioning the existence of any distinctive scientific ethos, they turned away from an idealised picture of consensus among scientists and instead became preoccupied with studying scientific debate and disagreement.

As a result of this shift in focus, many of what had been central issues in the classical sociology of science came to be generally neglected, among them issues relating to intellectual property and the openness of scientific communication. Moreover, because the new "sociology of scientific knowledge" incorporates a diverse range of theoretical approaches, its insights have not been as easily accessible to scholars outside sociology as those of Merton and his colleagues.

Today, most sociologists of science doubt the reality of normatively controlled behaviour, preferring instead to treat references to norms in the course of scientific debate as either mere rhetorical tools or rationalisations for interest driven behaviour. Nevertheless, the idea of scientific norms remained influential both within and outside the discipline of sociology, and when questions about the impact of commercialisation on scientific research attracted the attention of intellectual property lawyers in the late 1980s, possible effects on the normative structure of science provided a natural starting point for discussion.

DATASTREAM APPROACH

In an invited paper at the 1996 research tools workshop, sociologist Stephen Hilgartner moved to correct the neglect of issues relating to intellectual property and the role of scienctific communication by presenting a new model of scientific exchange--the "data stream" model, developed in a 1994 paper co-authored by Sherry Brandt-Rauf--incorporating insights from the new sociology of science. The data stream model is described here in some detail because it is highly relevant to the question of how legal aspects of commercialisation affect patterns of access to scientific research tools.

The first insight from the new sociology of science to inform the data stream model is that the concept of "data" should be subjected to social analysis rather than treated in commonsense terms. In contrast to earlier models of scientific research, the data stream model conceptualises data not as well-defined, stable entities--the end products of research--but as elements of an evolving data stream. Data streams have four key characteristics. First, they are composed of heterogeneous networks of information and resources, including many categories commonly used by scientists to describe the input and output of their work: data, findings, results, samples, materials, reagents, laboratory techniques, protocols, know-how, experience, algorithms, software and instruments. However, because the meaning of each of these terms is context-dependent, and each element is linked to many others in the data stream, it may be difficult to assign any given element to a single category.

Second, their elements range from mundane items which are part of the ordinary social infrastructure, such as water, electricity and computers, through elements specific to a research area but widely available either free or as commercial products, such as journal articles or assay kits, to specialised elements which are not publicly available but may be disseminated through personal contacts, and finally to novel or scarce elements available only by special arrangements. Hilgartner and Brandt-Rauf remark that critical issues in the analysis of scientific access practices most often concern elements of the data stream lying towards the "novel or scarce" end of this spectrum.

The third property of data streams is that different elements have different information status. At one extreme, elements of a data stream may be generally accepted as reliable and valuable, while at the other, they may be so uncertain that even the scientists who produce them doubt their credibility or usefulness. Data are constantly interpreted and reinterpreted through the research process, so that scientists' perceptions of the reliability and value of particular parts of the data stream vary with time; this can be important in decisions about access, as scientists ask themselves whether data are "ready" for dissemination, or how much data are "worth".

Finally, data streams are composed of chains of products. Scientists initially record data using primary inscription devices, such as x-ray film or electrophoresis gel, then convert the data into second, third or fourth order inscriptions; materials may processed and purified; electronic information may be subjected to a series of manipulations, and so on. Hilgartner and Brandt-Rauf argue that these translations and conversions affect access practices because they alter not only the information content and material form of the data, but also the purposes for which they can be used.

The second insight which Hilgartner and Brandt-Rauf draw from recent social studies of science is that transactions involving data are negotiated within complex research networks. They argue that analyses of data access patterns are often framed in terms of relationships between two parties--the primary researcher or producer of the data and the secondary researcher who wants to obtain access--but that in reality, each member of a research network is linked to many other people and organisations. Moreover, access practices are intimately involved in the construction and maintenance of such networks.

Therefore, the analysis of data access practices should take account of a range of relevant actors. A decision about whether to grant access to data may involve many parties: a research team of scientists, possibly from several institutions or several fields of study, with different levels of training and of involvement in the project; government and corporate sponsors providing funds; perhaps also a host university, with all its internal bureaucracy. These parties may have different goals and differing claims to portions of the data stream, and they may disagree about the optimal means and timing of dissemination. Similarly, audiences or markets for data do not necessarily consist of individuals or undifferentiated groups: they may include competing research groups, potential collaborators, authors of studies with conflicting results, gatekeepers who control key resources (e.g. department heads, corporate sponsors), potential markets for research based products, or venture capitalists.

The third relevant insight from the new sociology of science is that there is a wide range of mechanisms available for granting, limiting or denying access to data, and that analysis of data access practices should take into account the incentives and strategic considerations associated with each. While traditional models of data access emphasise peer recognition as a scientist's primary reward for discovery, with open publication as the primary legitimate means of achieving recognition, open publication is only one of many mechanisms for disseminating portions of a data stream. Data may be bartered in negotiations with prospective collaborators or sponsors, distributed to selected colleagues, patented, transferred by visitors being trained in new techniques, provided to a limited group on a confidential basis, bought and sold, pre-released to existing sponsors, kept in the lab pending future decisions about disposition, and so on. Hilgartner and Brandt-Rauf again identify a spectrum, from limited access to widespread distribution, and argue that as access becomes more widespread, the competitive edge conferred by possession of unique data declines.

Scientists can exploit this competitive edge by restricting access, using data to produce more data, or by providing carefully targeted access; or they may choose to provide widespread access in order to enhance their scientific reputation. Other factors affecting decisions about how to provide access include timing, the portion of the data stream to be made available, and the costs and logistics associated with different modes of access. Hilgartner and Brandt-Rauf observe that in order to comprehend these strategic issues in relation to a particular area of research, it is necessary to acquire a detailed understanding of the structure of data streams in that area.

The fourth and final insight from the new sociology of science is to examine how access practices interact with strategies for commercialisation. In his 1994 article with Brandt-Rauf and more recently, Hilgartner acknowledges that legal mechanisms of commercialisation may have a significant impact on scientific data access practices, noting that the law is clearly relevant to these practices because it addresses questions of ownership and control. He describes the legal approach to data ownership as atomistic: it involves plucking items from the data stream and attempting to place them into discrete categories in order to designate an end product that may qualify for some type of protection--patent, copyright, trade secrets, misappropriation, contract, or conversion (fn:transferred material is bailed property)-- while data which are not construed as falling into one of these categories are considered to fall within the public domain. Distinguishing between areas in which the law offers a relatively stable set of data protection mechanisms, and areas in which the law is still evolving (so that legal and scientific practices are simultaneously constructed in part through their interaction), Hilgartner proposes that future empirical research should focus on the relationship between scientific practices and the law. In particular, he believes it is important to understand how researchers try to employ legal mechanisms for controlling data access, what dilemmas and strategies are created by the disparity between the law's reductionist approach to ownership and the continuity of data streams and research networks, and how access practices, the law and the political economy of research interact to redefine legal regimes governing fast-moving areas such as biotechnology.

Applying the data stream perspective to the issue of whether an emphasis on intellectual property in academic science should be expected to cause a reduction in scientific openness, Hilgartner suggests we need answers to a series of empirical questions. Do intellectual property considerations influence what portions of data streams are provided, to whom, and when? Do they introduce new sources of delay, or change the kinds of restrictions that are placed on the use of data? Do intellectual property considerations increase the complexity and formality of negotiations over access to data, make collaborations more unstable or difficult to form, or complicate the development and maintenance of shared understandings about control over data streams that are collectively produced? Hilgartner argues that at any rate, we should not expect intellectual property protection to lead to an increase in openness among academic scientists: restrictions on openness motivated by possible commercial exploitation probably tend to propagate upstream from the point of potential commercialisation back into the research process, so that portions of data streams that are believed to be precursors of potentially patentable products are likely to be relatively tightly controlled. He notes that existing empirical evidence suggests intellectual property considerations do actually reduce openness, but warns that the effects of intellectual property protection on academic science will not be uniform across all fields, hypothesising that access practices are most intensively shaped not at the level of the discipline or field but at levels of research that can be defined in terms of a characteristic data stream and a particular competitive structure.

Ultimately, in Hilgartner's view, the most important questions about scientific data access practices are normative. Like Eisenberg responding to the concerns of research scientists in the 1980s, Hilgartner asks whether the public domain should be defended against encroachment by proprietary categories of information--though in the light of sociological literature problematising the concepts of "public" and "private" in scientific research, we should perhaps prefer his alternate formulation, in which the problem is expressed as one of deciding which data access policies are most likely to contribute to research productivity while promoting other social goals. Hilgartner's emphasis on the policy implications of data access practices is consistent with Eisenberg's 1987 argument that intellectual property policy-makers should be interested in the norms of science because legal rules are more likely to achieve their desired effect if they resonate with scientists' existing conceptions of appropriate behaviour. Not only does Hilgartner's empirical work demonstrate the continuing relevance of access issues in research science, his theoretical approach is particularly valuable from a legal policy perspective because, unlike earlier theories of scientific exchange, it confronts the complexity of scientific research systems.

This complexity is reflected not only in the problems faced by policy-makers, but also in the wide range of techniques which have been used to try to influence access practices, from informal admonishment, through formal rules imposed on grantees by funding agencies to reforming national and international intellectual property laws, to in-built technical solutions such as those employed in relation to human genome research. Hilgartner makes the important point that there are limits to what can be achieved by these methods: no single policy can resolve all the problems surrounding access to data in a particular field, because access issues reflect tensions--competition versus cooperation, "public" versus "private"--that are deeply embedded in the structure of scientific research. Nevertheless, policy-makers must do their best to prevent and resolve disputes over access because, regardless of where the merits lie, such disputes interfere with research by diverting resources away from the central business of knowledge production.

One limitation of Hilgartner's theoretical approach is worth mentioning. Eisenberg's argument in favour of designing intellectual property laws with an eye to existing scientific incentives has recently been picked up by Robert Merges and Arti Kaur Rai, writing from a law and norms perspective. Though promising, these analyses suffer from the necessity of drawing on the classical sociology of science literature for their understanding of scientific norms, even while acknowledging modern criticism of that literature. Hilgartner's focus on data access practices represents a move towards a more sophisticated understanding of the nature of these norms in the light of recent advances in the sociology of science, but there is still a gap between old and new approaches to the social study of scientific research; until these approaches have been reconciled as far as possible, we risk losing much that is still valuable in the classical sociology of science. There are signs of hope in Hilgartner's writing: he acknowledges that the data stream model does not take sufficient account of rhetoric based on traditional scientific norms, which may be important in shaping access negotiations, and points to the need to explain how collective definitions of appropriate conduct in science influence access practices.

Aside from this limitation, Hilgartner's data stream model provides a sound sociological basis for analysing the relationship between intellectual property law (and other legal mechanisms for controlling access to data) and data access practices in areas of scientific research undergoing rapid commercialisation, like biotechnology.

NORMATIVE STRUCTURE

The implication of this is that there are arbitrary characteristics of the legal structure, and as a result normative characteristics of the society as a whole, which depend on which kind of system the societies members believe the normative structure has turned out to be. As we can see, this depends on how the members of the society see themselves in relationship to the creation and maintenance of their laws. This does not necessarily mean that the members of the society must actively or physically participate in the choice of norms. In a theocracy for example members of the society have no hand in the determination of the norm structure yet because they have an inner feeling that the norms chosen are those they agree with, they feel that they are part of the system. At the same time, in a democracy strong minority groups can influence legislation that will alienate the majority while bigoted majority opinions will inevitably alienate some minorities. It is also Important to realize that these images do not have to be real, they need only be the consensus of the prevailing beliefs of the society, or in extreme cases Just those members of the society who are affected.

Social systems which do not serve the society they develop within will not continue to exist. At the same time, the society as a whole emerges out of the Interaction of people who act through social systems. The purpose of a legal system, as we have gone over before, is to provide a congruence of social norms. The laws of a society make up one level of the complex legal system. A level that occurs at the interface between the serial Interactions of members playing out those roles which involve the legal system and the parallel communications media that exist between members and which is relative to legal problems, by these we mean problems relating to normative disappointments.

When we say along with Luhmann that through legal structures norms are counter factually stabilized we mean that the norms fixed by the legal system have no non-social Justification for their existence that would not be Just as applicable to norms which have not been selected. Thus the only validity that legal norms have is that they can be changed but have not been.

Therefore, the only criteria an individual has for assessing the validity of a particular law is his own personal commitment to Its longevity. Such commitment is derived from the responsibility he feels for its enactment, either through his own activity or through some other who he feels speaks with a sympathetic voice and is in some responsible to him. It also depends on the faith he has that should the law prove to be detrimental to his own Inner beliefs he, through some agency he has faith in, could have it changed or restricted. Since no law could ever stand such scrutiny In more than a very limited society, all laws require some higher constraint. This must choose from among all possible Interpretations those which will result In the best consensus among those who are subject to the legal system. This level corresponds to Luhmann's level of values.

THE EMERGENCE OF PRINCIPLES

There are two criteria for a level of legal activity which can Justifiably constrain the exercise of laws. One is that it must emerge out of the incongruencies that arise out of the administration of laws, and two, It must deal with the application of laws and not with their creation. In modern positive law that level is Judicial review and its mechanism is legal precedence.

These limit the interpretation of laws to standards that have already been tested and determined to be acceptable. Thus every questionable interpretation of a law doesn't have to be re-fought every time it comes up. Judicial review is a higher level of the legal system constraining the administration of the laws themselves to maintain stability in the face of overwhelming varieties of legal interpretation. However, review alone is not sufficient. It deals with specific Interpretations about the ways that laws deal with specific classes of cases. Judicial review itself must be constrained by a level with even greater abstraction, one that corresponds to Luhmann's level of social values, this is the level that corresponds to Dworkin's theory of legal principles.

It might seem that as I discuss Dworkin's concept of principles there is a point of opposition between his ideas and mine. Arguing from the view of an essentially Just society, his two principles appear to derive from some natural law while analyzing the structure of legal systems from a sociological point of view I arrive at descriptions which will include his but only as one possibility out of many. But keep In mind that Dworkin has developed his principles out of a study of the way Judges in actual essentially Just societies make decisions. Part of my contention is that a society will be a Just society only if those who are responsible for enacting its laws want it to. Thus a Just society is one possible expression of normative structure which a society may choose, but need not. However, if the normative structure is seen as a complex system having the property that the higher levels emerge from and at the same time constrain the lower then it can be shown that a more just society is more likely to be more stable than one which is less Just.

In "A matter of Principle" Dworkin argues effectively that his principles actually do express the actions of Judges involved in Judicial review better than those alternatives which have been attempted by others. What I want to show is the relationship between these two principles and my theory of complex structure. Let us begin with the principles themselves. First, any political decision must treat all citizens as equals, that is, as equally entitled to concern and respect. Second, if a political decision is taken and announced that respects equality as demanded by the first principle then a later enforcement of that decision Is not a fresh political decision that must also be equal In Its Impact In that Way.

The first principle seems to lead us Into controversy since the meaning of equality can be taken In any number of ways. For example, in one society where the majority of the people feel that a minority group within the society are incapable of deciding moral Issues on their own, making such decisions for them would be considered treating them with concern and respect. In another society this might be taken also to mean that all persons in the society must be maintained at relatively equal economic levels. These issues do not concern Dworkin because he is not developing a theory of Justice. He is concerned with the principles Judges use when they make legitimate political Judgments. That means that assuming any set of standards a society may adopt for themselves, the laws that they pass must be Judged as they apply to the citizens of the society. If they apply equally to all then they are legitimate, If not then they are Illegitimate.

MORAL HARM

The difference between Dworkin's conception of equality and others lies In his theory of moral harm. Considering the first principle he says, "It is part of the principle that no decision may deliberately impose on any citizen a much greater risk of moral harm than it imposes on any other." For example let us take the right not to be convicted if innocent. According to Dworkin, the main function of this kind of right is as an instruction to the government. However, even if government heeds the instruction yet still makes a blameless mistake we should still consider government wrong because the violation of a right constitutes a special kind of harm. It is a harm that people suffer even when the violation is accidental Dworkin summed It up in these words.

We must distinguish, that Is, between what we might call the bare harm a person suffers through punishment, whether that punishment is just or unjust,--for example, the suffering or frustration or pain or dissatisfaction of desires that he suffers Just because he loses his liberty or is beaten or is killed--and the further

Injury that he might be said to suffer whenever his punishment Is unjust, just In virtue of that unjustice.

An Important point about moral harm is that it occurs even when no one is aware of it, in fact it occurs even when no one, or at least a very few care. This makes it a powerful concept. "Suppose," Dworkin argued, "we discover that some person executed for murder several decades ago was in fact innocent. We shall want to say that the world has gone worse than we thought, though we may add, if we reject the idea of moral harm, that no one suffered any harm of which we were ignorant, or was In any May worse off than we believed." Moral harm then, applies as much to situations as it does to people. Consider a society which enacts a set of laws that apply equally to all of its citizens. Some may disagree about the wisdom of those laws, but moral harm has nothing to do with the way people feel about laws, it only deals with the equality with which the rules are enacted and enforced.

This bring us to Dworkin's second principle. Dworkin amplified the concept in this way, "The second principle appeals to the fairness of abiding by open commitments fair when adopted--the fairness, for example, of abiding by the result of a coin toss when both parties reasonably agreed to the toss." For example, If we examine the deliberate conviction of someone known to be innocent from the standpoint of these two principles He see that this is a worse case of moral harm than that of a mistaken conviction. "Framing someone is a case of a fresh political decision that does not treat him as an equal as required by the first principle" This is not a case of the application of public commitments. Even if there were a public commitment to frame people who meet a specific test, this would not meet the requirements of the first principle. As Dworkin emphasized, the deliberate conviction of some one known to be innocent involves greater moral harm because it violates the equal standing of all citizens.

Legal principles emerge as a response to the failure of laws as they are written to apply directly to every case. Those that are successful at developing reactions to such disappointment that are generally accepted by the community will remain. Thus, from a study of the relatively Just societies of the United States and Britain Dworkin was able to Identify a pair of principles that he has extracted out of the more successful Judicial decisions. Their success, however, cannot be attributed to some natural or god-given concepts. They emerge naturally at the Interface between the serial network of Judicial review and the parallel network of the societies sense of justice. Cultures with very different concepts of Justice would develop different sets of principles. Thus principles are the highest level of the legal system. In other words, above that level the legal system is Just one element in a higher level system which emerges out of the interactions between it and other social systems as the cultures unique response to the universe.

THE SOCIOLOGY OF LAW

Let us return now to our sociological view. The normative structure of a society is hierarchical, the four levels as we have abstracted them are roles, laws, precedence, and principles. Each level constrains the freedom of the level below in order to maintain its own stability. At the same time each emerges from the activities of the level below. Unlike structures below that of the social, each level consists entirely of the actions of people and the reasons people accept for those actions. They all exist in the Inner worlds that members of the society create out of their interaction with their environment. The system of laws chooses for members of the society which of all of the possible norms that could be used will apply to their normative interaction with each other. They are counter factually stabilized because there exists no objective criteria why one set of norms should be selected rather than another that does not originate within the minds of the citizens.

The rules for the enactment and modification of laws are part of the system, thus, they tend to reflect the way members of the society view their norms. However, societies are not homogeneous, there are differing ways that Individuals interpret laws. Of course it is fortunate that this is so because otherwise in the case of an unexpected change in the kinds of problems that the individuals sense from their environment there might not be included in the variety of approaches available within the system one which would be successful in the changed environment. The system of laws exists at the interface between two different kinds of networks. The first is the serial network of individual normative disappointments. The second is a parallel neural network of congruency communications. Communications media exists wherever the manner of one partners selection serves simultaneously as the motivating structure for another. Stability is achieved when the norms communicated by the media result in input levels which exceed the threshold levels which define the norms that individuals have developed through serial interaction.

This paper Is not meant to be a treatise on a sense of justice, but a few characteristics of such a society will suffice to show that it does not have to be the only possible social arrangement. In the first place Dworkin stated that the best environment for his two principles is a majoritarian society. In such a society there need not be a sense of Justice, but, as William Nelson would concur, there is more likely to be one than in a society that is not. However, that is an act of faith. A sense of Justice is an inner conviction that exists in the mental world of the individuals in the society and not in the set of rules for deciding Issues.

On the other hand, in a society which lacks a sense of Justice that mirrors that of Dworkin, some other set of principles would be chosen by members of the society which would express their inner convictions. For example the minority governments of Ireland, Lebanon, and South Africa display a sense of Justice which assumes a class of inferior citizens. As a result the countries are constantly In turmoil. In the case of America except for the poor, or in England except in the case of Irish Catholics, Dworkin's two principles of Justice do apply. And, to the extent that they do, there is a stability in these societies.

CONCLUSION

By applying theories developed out of the study of complex systems we have succeeded in showing how legal systems evolve out of the failure of simple normative systems to maintain consensus when faced with an environment rich in excess variety. We see that laws evolve from normative disappointments, precedence from the failure of laws to pertain to specific cases, and principles from the failure of precedences to discriminate clearly between law and precedence.

Thus we come to the conclusion of our look Into the evolution of legal systems. Luhmann's concepts of differentiation and self-reflexivity, complimented by the complex concepts of hierarchical order and constraint meet the three requirements for an evolutionary system; a source of variety, a mechanism for selecting from among the variety, and a mechanism for fixing the choice. Luhmann's description of positive law is law which gains it validity from the fact that it can be changed. His concept of congruency communications becomes the criteria for the validation of the legal system. Finally, principles are the highest level of the legal system. Above that level the legal system Is just one element In a higher level system which emerges out of the interactions between it and other social systems as the cultures unique response to the universe.

HISTORY OF ANATOMY

The history of anatomy as a science extends from the earliest examinations of sacrificial victims to the sophisticated analyses of the body performed by modern scientists. It has been marked, over time, by a continually developing understanding of the functions of organs and structures in the body. Methods have also advanced dramatically, advancing from examination of animals through dissection of cadavers to technologically complex techniques developed in the 20th century.

Ancient Anatomy

Egypt

The study of anatomy begins at least as early as 1600 BCE, the date of the ancient Egyptian Edwin Smith papyrus. This treatise shows that the heart, its vessels, liver, spleen, kidneys, uterus and bladder were recognized, and that the blood vessels were known to come from the heart. Other vessels are described, some carrying air, some mucus, while two to the right ear are said to carry the "breath of life", and two to the left ear the "breath of death". The Ebers papyrus (c. 1550 BCE) features a treatise on the heart. It notes that the heart is the center of the blood supply, with vessels attached for every member of the body. The Egyptians seem to have known little about the function of the kidneys and made the heart the meeting point of a number of vessels which carried all the fluids of the body-blood, tears, urine and sperm.

Greece

The earliest medical scientist of whose works any great part survives today is Hippocrates, a Greek physician active in the late 5th and early 4th centuries BCE (460-377 BCE). His work demonstrates a basic understanding of musculoskeletal structure, and the beginnings of understanding of the function of certain organs, such as the kidneys. Much of his work, however, and much of that of his students and followers later, relies on speculation rather than empirical observation of the body.

In the 4th century BCE, Aristotle and several contemporaries produced a more empirically founded system, based on dissection of animals. Works produced around this time are the first to identify the difference between arteries and veins, and the relations between organs are described more accurately than in previous works.

The first use of human cadavers for anatomical research occurred later in the 4th century BCE when Herophilos and Erasistratus performed dissections of cadavers in Alexandria under the auspices of the Ptolemaic dynasty. Herophilos in particular developed a body of anatomical knowledge much more informed by the actual structure of the human body than previous works had been.

Galen

The final major anatomist of ancient times was Galen, active in the 2nd century. He compiled much of the knowledge obtained by previous writers, and furthered the inquiry into the function of organs by performing vivisection on animals. His collection of drawings, based mostly on dog anatomy, became the anatomy textbook for 1500 years. The original text is long gone, and his work was only known to the Renaissance doctors through the careful custody of Arabic medicine, since the Roman Catholic Church destroyed the work as heresy. Hampered by the same religious restrictions as anatomists for centuries after him, Galen assumed that anatomical structures in dogs were the same as for humans.

Medieval to Early Modern Anatomy

Little progress was made in anatomy for some time after the fall of the Roman empire. Although Arab scientists contributed heavily to medieval learning and culture, taboos against handling corpses limited their contributions to anatomical research. The first major development in anatomy after Galen occurred at Bologna in the 14th to 16th centuries, where a series of authors dissected cadavers and further contributed to the accurate description of organs and the identification of their functions. Prominent among these anatomists were Mondino de Liuzzi and Alessandro Achillini.

The 16th century also saw the first challenges to Galen. Thanks to the printing press, all over Europe a collective effort proceeded to distill the original Galen from the various, mostly Arab, added texts. Vesalius was the first to publish a treatise that challenged him "drawing for drawing", travelling all the way from Leuven to Padua for permission to dissect victims from the gallows without fear of persecution.

His drawings are triumphant descriptions of the, sometimes major, descrepancies between dogs and humans, showing superb drawing ability. Many later anatomists challenged Galen in their texts, though Galen reigned supreme for another century.

A succession of researchers proceeded to further refine the body of anatomical knowledge, giving their names to a number of anatomical structures along the way. The 16th and 17th centuries also witnessed significant advances in the understanding of the circulatory system, as the purpose of valves in veins was identified, the left-to-right ventricle flow of blood through the circulatory system was described, and the hepatic veins were identified as a separate portion of the circulatory system. The lymphatic system was also identified as a separate system at this time.

17th and 18th Century Anatomy

The study of anatomy flourished in the 17th and 18th centuries. With the coming of the printing press, the exchange of ideas across Europe was easily facilitated. Since the study of anatomy concerned observation and drawings, the popularity of the anatomist was equal to the quality of his drawing talents, and one needn't be an expert in Latin to take part.

Many famous artists studied anatomy, attended dissections, and published drawings for money, from Michelangelo to Rembrandt. For the first time, prominent universities could teach something about anatomy through drawings, rather than relying on knowledge of Latin. The only stumbling block was a possible reprimand from the Church, which frightened several anatomists of that time from performing dissections on their own kind. Though a very fruitful period for the sciences, the Renaissance could be dangerous, as is seen in the case of Galileo. Some scientists of the time were scared enough to keep moving from city to city. Descartes is a prime example. Though all physicians agreed that a sound knowledge of anatomy was important to perform medicine, only certified anatomists were allowed to perform dissections, sometimes only yearly. These dissections were sponsored by the city councilors and often charged an admission fee, rather like a circus act for scholars. Many European cities, such as Amsterdam, London, Copenhagen, Padua, and Paris, all had Royal anatomists (or some such office) tied to local government. Indeed, Nicolaes Tulp was Mayor of Amsterdam for three terms. Though it was a risky business to perform dissections, attending dissections was perfectly legal, and many anatomy students traveled around Europe from dissection to dissection during the course of their study.

Many Europeans interested in the study of anatomy traveled to Italy, where the center of anatomy resided. Only in Italy could certain important research methods, such as dissections on women, be used. M. R. Columbus and Gabriele Falloppio were pupils of Vesalius, the 16th century anatomist.

Columbus, as his immediate successor in Padua, and afterwards professor at Rome, distinguished himself by rectifying and improving the anatomy of the bones; by giving correct accounts of the shape and cavities of the heart, of the pulmonary artery and aorta and their valves, and tracing the course of the blood from the right to the left side of the heart; by a good description of the brain and its vessels, and by correct understanding of the internal ear, and the first good account of the ventricles of the larynx. Osteology at nearly the same time found an assiduous cultivator in Giovanni Filippo Ingrassias.

19th Century Anatomy

The 19th century saw anatomists largely finalise and systematise the descriptive human anatomy of the previous century. The discipline also progressed to establish growing sources of knowledge in histology and developmental biology, not only of humans but also of animals. Extensive research took place in a growing number of areas, with England being a particular center of research.

Demand for cadavers grew so great there that body-snatching and even murder came into use as a means of obtaining them. In response, parliament passed the Anatomy Act 1832, which finally provided for an adequate and legitimate supply of corpses. The relaxed restrictions on dissection provided the groundwork for Gray's Anatomy, a text that was a collective effort and became widely popular. Now seen as unwieldy, Gray's Anatomy was born out of a need to create a single volume on anatomy for the traveling doctor.

MODERN ANATOMY

Anatomical research in the past hundred years has taken advantage of technological developments and growing understanding of sciences such as evolutionary and molecular biology to create a thorough understanding of the body's organs and structures. While disciplines such as endocrinology have explained the purpose of glands that previous anatomists could not explain, medical devices such as MRI machines and CAT scanners have enabled researchers to study the organs of living people. Progress today in anatomy is centered in the field of molecular biology, as the macroscopic aspects of the field have now been catalogued and addressed.

ANECDOTAL COGNITIVISM

Anecdotal cognitivism is a psychological theory and animal cognition term which entails attribution of mental states to animals on the basis of anecdotes, and on the observation of particular cases, other than that of controlled experiments. It is opposed to behaviorism, and behaviorists are very critical of anecdotal cognitivism, suggesting that controlled experiments are necessary to correctly measure stimuli and record observable behavior.

Anecdotal cognitivism is often criticized by behaviorists using specific cases, such as that of Clever Hans, to discredit using anecdotal evidence in assessing animal cognition. In the case of clever Hans, a horse was purported to be able to add and subtract using its hooves, and even answer questions surrounding European politics, but it was determined by later research that the horse's owner was in fact unknowingly cueing the horse, and that when he was removed from the room, the horse would not respond.

Anecdotal cognitivists respond to behaviorists by saying that behaviorism would have the animals 'lose their minds', and that it is clear that by observation we can know a great deal about the cognitive processes of animals, and that the debate can start here, with simple observation, and not in a controlled setting or in a lab.

5

HISTORY OF HUMAN ARTIFACTS

The idea of human artifacts being given life has fascinated mankind for as long as men have been recording their myths and stories. Whether Pygmalion or Frankenstein, mankind has been fascinated with the idea of artificial life.

Pre-computer

Automatons were (and arguably are) quite a novelty. In the days before computers and electronics, some were very sophisticated, using pneumatics, mechanics, and hydraulics. Most famous was an artificial duck, with thousands of moving parts, created by Jacques de Vaucanson. The duck could reportedly eat and digest, drink, quack, and splash in a pool. It was exhibited all over Europe until it fell into disrepair.

However, it wasn't until the invention of cheap computing power that artificial life as a legitimate science began in earnest, steeped more in the theoretical and computational than the mechanical and mythological.

1950s-1970s

One of the earliest thinkers of the modern age to postulate the potentials of artificial life, separate from artificial intelligence, was math and computer prodigy John Von Neumann. At the Hixon Symposium, hosted by Linus Pauling in Pasadena, California in the late 1940s, Von Neumann delivered a lecture titled "The General and Logical Theory of Automata." He defined an "automaton" as any machine whose behavior proceeded logically from step to step by combining information from the environment and its own programming, and said that natural organisms would in the end be found to follow similar simple rules. He also spoke about the idea of self-replicating machines. He postulated a machine--a kinematic automaton--made up of a control computer, a construction arm, and a long series of instructions, floating in a lake of parts. By following the instructions that were part of its own body, it could create an identical machine.

He followed this idea by creating (with Stanislaw Ulam) a purely logic-based automata, not requiring a physical body but based on the changing states of the cells in an infinite grid--the first cellular automaton. It was extraordinarily complicated compared to later CAs, having hundreds of thousands of cells which could each exist in one of twenty-nine states, but Von Neumann felt he needed the complexity in order for it to function not just as a self-replicating "machine", but also as a universal computer as defined by Alan Turing. This "universal constructor" read from a tape of instructions and wrote out a series of cells that could then be made active to leave a fully functional copy of the original machine and its tape. Von Neumann worked on his automata theory intensively right up to his death, and considered it his most important work.

Homer Jacobson illustrated basic self-replication in the 1950s with a model train set--a seed "organism" consisting of a "head" and "tail" boxcar could use the simple rules of the system to consistently create new "organisms" identical to itself, so long as there was a random pool of new boxcars to draw from. Edward F. Moore proposed "Artificial Living Plants", which would be floating factories which could create copies of themselves. They could be programmed to perform some function (extracting fresh water, harvesting minerals from seawater) for an investment that would be relatively small compared to the huge returns from the exponentially growing numbers of factories. Freeman Dyson also studied the idea, envisioning self-replicating machines sent to explore and exploit other planets and moons, and a NASA group called the Self-Replicating Systems Concept Team performed a 1980 study on the feasibility of a self-building lunar factory.

University of Cambridge professor John Horton Conway invented the most famous cellular automaton in the 1960s. He called it the Game of Life, and publicized it through Martin Gardner's column in Scientific American magazine.

1970s-1980s

Philosophy scholar Arthur Burks, who had worked with Von Neumann (and indeed, organized his papers after his death), headed the Logic of Computers Group at the University of Michigan. He brought the overlooked views of 19th century American thinker Charles S. Peirce into the modern age. Peirce was a strong believer that all of nature's workings were based on logic.

The Michigan group was one of the few groups still interested in alife and CAs in the early 1970s; one of its students, Tommaso Toffoli argued in his PhD thesis that the field should not be overlooked as a mathematical curiosity, because its results were so powerful in explaining the simple rules that underlay complex effects in nature. Toffoli later provided a key proof that CAs were reversible, just as the true universe is considered to be.

Christopher Langton was an unconventional researcher, with an undistinguished academic career that led him to a job programming DEC mainframes for a hospital. He became enthralled by Conway's Game of Life, and began pursuing the idea that the computer could emulate living creatures. After years of study (and a near-fatal hang-gliding accident), he began attempting to actualize Von Neumann's CA and the work of Edgar F. Codd, who had simplified Von Neumann's original twenty-nine state monster to one with only eight states. He succeeded in creating the first self-replicating computer organism in October of 1979, using only an Apple II desktop computer. He entered Burks' graduate program at the Logic of Computers Group in 1982, at the age of 33, and helped to found a new discipline.

Langton's official conference announcement of Artificial Life I was the earliest description of a field which had previously barely existed: Artificial life is the study of artificial systems that exhibit behavior characteristic of natural living systems. It is the quest to explain life in any of its possible manifestations, without restriction to the particular examples that have evolved on earth. This includes biological and chemical experiments, computer simulations, and purely theoretical endeavors. Processes occurring on molecular, social, and evolutionary scales are subject to investigation. The ultimate goal is to extract the logical form of living systems.

Microelectronic technology and genetic engineering will soon give us the capability to create new life forms in silico as well as in vitro, This capacity will present humanity with the most far-reaching technical, theoretical and ethical challenges it has ever confronted. The time seems appropriate for a gathering of those involved in attempts simulate or synthesize aspects of living systems.

Ed Fredkin founded the Information Mechanics Group at MIT, which united Toffoli, Norman Margolus, Gerard Vichniac, and Charles Bennett. This group created a computer especially designed to execute cellular automata, eventually reducing it to the size of a single circuit board. This "cellular automata machine" allowed an explosion of alife research among scientists who could not otherwise afford sophisticated computers.

In 1982, computer scientist Stephen Wolfram turned his attention to cellular automata. He explored and categorized the types of complexity displayed by one-dimensional CAs, and showed how they applied to natural phenomena such as the patterns of seashells and the nature of plant growth. Norman Packard, who worked with Wolfram at the Institute for Advanced Study, used CAs to simulate the growth of snowflakes, following very basic rules.

Computer animator Craig Reynolds similarly used three simple rules to create recognizable flocking behaviour in groups of computer-drawn "boids" in 1987. With no top-down programming at all, the boids produced life-like solutions to evading obstacles placed in their path. Computer animation has continued to be a key commercial driver of alife research as the creators of movies attempt to find more realistic and inexpensive ways to animate natural forms such as plant life, animal movement, hair growth, and complicated organic textures.

The Unit of Theoretical Behavioural Ecology at the Free University of Brussels applied the self-organization theories of Ilya Prigogine and the work of entomologist E.O. Wilson to research the behavior of social insects, particularly allelomimesis, in which an individual's actions are dictated by those of a neighbor.

They developed partial differential equations which modeled the shapes created by termites when constructing their nest. They then compared that to the reaction of real termites to identical changes in laboratory colonies, and refined their theories about the rules which underlay the behavior.

J. Doyne Farmer was a key figure in tying artificial life research to the emerging field of complex adaptive systems, working at the Center for Nonlinear Studies (a basic research section of Los Alamos National Laboratory), just as its star chaos theorist Mitchell Feigenbaum was leaving. Farmer and Norman Packard chaired a conference in May of 1985 called "Evolution, Games, and Learning", which was to presage many of the topics of later alife conferences.

Work is underway to create cellular models of artificial life. Initial work on building a complete biochemical model of cellular behavior is underway as part of a number of different research projects, namely BlueGene which seeks to understand the mechanisms behind protein folding.

ASILOMAR CONFERENCE ON RECOMBINANT DNA

The Asilomar Conference on Recombinant DNA was an influential conference organized by Paul Berg discussing the potential biohazards and regulation of biotechnology held in February 1975 at a conference center Asilomar State Beach. The Asilomar Conference was organized by leading researchers such as Paul Berg; A group of around 140 professionals (primarily biologists, but also including lawyers and physicians) participated in the conference to draw up voluntary guidelines to ensure the safety of recombinant DNA technology. The conference also placed scientific research more into the public domain, and can be seen as applying a version of the precautionary principle.

The repercussions of these actions are still being felt through the biotechnology industry and the participation of the general public in scientific discourse. Due to potential safety hazards, scientists worldwide had halted experiments using recombinant DNA technology, which entailed combining DNAs from different organisms2,3. After the establishment of the guidelines during the conference, scientists continued with their research, which increased fundamental knowledge about biology and the public's interest in biomedical research.

BACKGROUND: RECOMBINANT DNA TECHNOLOGY

Recombinant DNA technology arose as a result of advances in biology that began in the 1950's and 60's. During these decades, a tradition of merging the structural, biochemical and informational approaches to the central problems of classical genetics became more apparent. Two main underlying concepts of this tradition were that genes consisted of DNA and that DNA encoded information that determined the processes of replication and protein synthesis. These concepts were embodied in the model of DNA proposed by James Watson and Francis Crick, and further research on the Watson-Crick model yielded theoretical advances that were reflected in new capacities to manipulate DNA. One of these capacities was recombinant DNA technology.

EXPERIMENTAL DESIGN

This technology entails the joining of DNA from different species and the subsequent insertion of the hybrid DNA into a host cell. One of the first individuals to develop recombinant DNA technology was a biochemist at Stanford by the name of Paul Berg.

In his experimental design in 1972, he cleaved (cut into fragments) the monkey virus SV40. He then cleaved the double helix of another virus; an antibacterial agent known as bacteriophage lambda. In the third step, he fastened DNA from the SV40 to DNA from the bacteriophage lambda. The final step involved placing the mutant genetic material into a laboratory strain of the E. coli bacterium. This last step, however, was not completed in the original experiment.

INITIAL BIO-SAFETY CONCERNS

Berg did not complete his final step due to the pleas of several fellow investigators who feared the biohazards associated with the last step. The SV40 was known to cause cancer tumors to develop in mice. Additionally, the E. coli bacterium (although not the strain used by Berg) inhabited the human intestinal tract.

For these reasons, the other investigators feared that the final step would create cloned SV40 DNA that might escape into the environment and infect laboratory workers. These workers could then become cancer victims.

Concern about this potential biohazard, along with others, caused a group of leading researchers to send a letter to the president of the National Academy of Science (NAS). In this letter, they requested that he appoint an ad hoc committee to study the bio-safety ramifications of this new technology.

This committee, called the Committee on Recombinant DNA molecules of the National Academy of Science, U.S.A., held in 1974, concluded that an international conference was necessary to resolve the issue and that until that time, scientists should halt experiments involving recombinant DNA technology.

ESTABLISHED PRINCIPLES

The Asilomar Conference on Recombinant DNA took place at the Asilomar Conference Center on California's Monterey Peninsula in 1973. The main goal of the conference was to address the biohazards presented by recombinant DNA technology. During the conference, the principles guiding the recommendations for how to conduct experiments using this technology safely were established. The first principle for dealing with potential risks was that containment should be made an essential consideration in the experimental design. A second principle was that the effectiveness of the containment should match the estimated risk as closely as possible.

The conference also suggested the use of biological barriers to limit the spread of recombinant DNA. Such biological barriers included fastidious bacterial hosts that were unable to survive in natural environments. Other barriers were nontransmissible and equally fastidious vectors (plasmids, bacteriophages, or other viruses) that were able to grow in only specified hosts.

In addition to biological barriers, the conference advocated the use of additional safety factors. One such safety factor was physical containment, exemplified by the use of hoods or where applicable, limited access or negative pressure laboratories. Another factor was the strict adherence to good microbiological practices, which would limit the escape of organisms from the experimental situation. Additionally, the education and training of all personnel involved in the experiments would be essential to effective containment measures.

RECOMMENDATIONS

The Asilomar Conference also gave recommendations for matching the types of containment necessary for different types of experiments. These recommendations were based on the different levels of risk associated with the experiment, which would require different levels of containment. These levels were minimal, low, moderate and high risk. The minimal risk level of containment was intended for experiments in which the biohazards could be accurately assessed and were expected to be minimal.

Low risk containment was appropriate for experiments that generated novel biotypes but where the available information indicated that the recombinant DNA could not either alter appreciably the ecological behavior of the recipient species, increase significantly its pathogenicity or prevent effective treatments of any resulting infections. The moderate risk level of containment was intended for experiments in which there was a probability of generating an agent with a significant potential for pathogenicity or ecological disruption. High-risk containment was intended for experiments in which the potential for ecological disruption or pathogenicity of the modified organism could be severe and thereby pose a serious biohazard to laboratory personnel or to the public. These levels of containments, along with the previously mentioned safety measures, formed the basis for the guidelines used by investigators in future experiments that involved the construction and propagation of recombinant DNA molecules using DNA from prokaryotes, bacteriophages and other plasmids, animal viruses and eukaryotes.

RECOMMENDATIONS APPLIED TO EXPERIMENTS

For prokaryotes, bacteriophages and other plasmids, experiments could be performed in minimal risk containment facilities when the construction of recombinant DNA molecules and their propagation involved prokaryotic agents that were known to exchange genetic information naturally. For experiments involving the creation and propagation of recombinant DNA molecules from DNAs of species that ordinarily did not exchange genetic information and generate novel biotypes, the experiments were to be performed in at least in a low risk containment facility. If the experiment increased the pathogenicity of the recipient species or result in new metabolic pathways in species, then moderate or high-risk containment facilities were to be used. In experiments where the range of resistance of established human pathogens to therapeutically useful antibiotics or disinfectants was extended, the experiments were to be undertaken only in moderate or high-risk containment facilities.

When working with animal viruses, experiments that involved the linkage of viral genomes or genome segments to prokaryotic vectors and their propagation in prokaryotic cells were to be conducted only with vector-host systems that had demonstrated restricted growth capabilities outside the laboratory and in moderate risk containment facilities. As safer vector-host systems became available, such experiments could be performed in low risk facilities. In experiments designed to introduce or propagate DNA from non-viral or other low risk agents in animal cells, only low risk animal DNA could be used as vectors and the manipulations were to be confined to moderate risk containment facilities.

With eukaryotes, attempts to clone segments of DNA using recombinant DNA technology from warm-blooded vertebrates genomes were to be performed only with vector-host systems that had demonstrably restricted growth capabilities outside the laboratory and in a moderate risk containment facility. This was because they potentially contained cryptic viral genomes that were potentially pathogenic to humans However, unless the organism made a dangerous product, recombinant DNAs from cold-blooded vertebrates and all other lower eukaryotes could be constructed and propagated with the safest vector-host system available in low risk containment facilities. Additionally, purified DNA from any source that performed known functions and was judged to be non-toxic could be cloned with available vectors in low risk containment facilities.

PROHIBITED EXPERIMENTS

In addition to regulating the experiments that were conducted, the guidelines also forbade the performance of other experiments. One such experiment was the cloning of recombinant DNAs derived from highly pathogenic organisms. In addition, neither the cloning of DNA containing toxin genes, nor large scale experiments using recombinant DNAs that were able to make products that were potentially harmful to man, animals or plants were allowed under the guidelines. These experiments were banned because the potential biohazards could not be contained by the then current safety precautions.

SCIENCE AND THE GENERAL PUBLIC

The participants of the Asilomar Conference also endeavored to bring science into the domain of the general public, with a possible motivation being the Watergate scandal. The scandal resulted from a bungled break-in at the Watergate hotel, which served as the Democratic National Committee headquarters in 1972. Two years after the burglary, taped evidence was discovered that indicated that President Nixon had discussed a cover-up a week after it.

Three days following the release of the tape, Nixon resigned from his presidential office.

This event focused the nation's attention on the problem of government secrecy fostering illegal and immoral behavior and it has been suggested by the political scientist, Ira H. Carmen, that this motivated the scientists at the Asilomar Conference to bring science into the public eye to ensure that they would not be accused of a cover-up.. Additionally, according to Dr. Berg and Dr. Singer, by being forthright, scientists avoided restrictive legislation due to the development of a general consensus on how they were to conduct their research.

Bringing science into the public eye also coincided with the rapid rate at which recombinant DNA technology entered the industrial world. Because of the practical applications of the technology, funding for research using it started coming more from the private sector and less from the public sector. In addition, many molecular biologists who once confined themselves to academia, developed ties with private industry as equity owners, corporate executives and consultants.

This led to the creation of a biotechnology industry, although during this time, public debates occurs over the hazards of recombinant DNA. These debates were eventually won over by scientists who stated that the hazards were exaggerated and that the research could be conducted safely.

Such was seen in the Ascot report, found in the Federal Register in March of 1978. This report emphasized that the hazards of recombinant DNA to the general community were small to the point that they were of no practical consequence to the general public.

For this reason, along with high economic pressures for industrial development and a more supportive political environment that existed after 1979, research and industry based on recombinant DNA continued to expand.

SIGNIFICANCE OF THE CONFERENCE

Years after the conference, people ascribed a large amount of significance to it. According to Paul Berg and Maxine F. Singer in 1995, the conference marked the beginning of an exceptional era for both science and the public discussion of science policy.

The guidelines devised by the conference enabled scientists to conduct experiments with recombinant DNA technology, which by 1995 dominated biological research. This research, in turn, increased knowledge about fundamental life processes, such as the cell cycle.

Additionally, the conference along with public debates on recombinant DNA, increased public interest in biomedical research and molecular genetics. For this reason, by 1995, genetics and its vocabulary had become a part of the daily press and television news. This, in turn, stimulated knowledgeable public discussion about some of the social, political and environmental issues that emerged from genetic medicine and the use of genetically modified plants in agriculture.

Another significant outcome of the conference was the precedent it set about how to respond to changes in scientific knowledge. According to the conference, the proper response to new scientific knowledge was to develop guidelines that governed how to regulate it.

6

GENETIC ENGINEERING AND BIOLOGY

Biotechnology is the application of scientific and engineering principles to the processing of materials by biological agents to provide goods and services. From its inception, biotechnology has maintained a close relationship with society. Although now most often associated with the development of remarkable drugs, historically biotechnology has been principally associated with food, addressing such issues as malnutrition and famine. The history of biotechnology begins with zymotechnology, which commenced with a focus on brewing techniques for beer. By World War I, however, zymotechnology would expand to tackle larger industrial issues, and the potential of industrial fermentation gave rise to biotechnology.

The belief that the needs of a society could be met by fermenting industrial waste was an important ingredient of the "chemurgic movement," which began in the 1930s. Expectations were raised during the 1960s by a process that grew single-cell protein from petroleum oil. This new product was seen to be a solution to growing food shortages and world hunger. In the 1970s, biotechnology offered a possible solution to the escalation in oil prices and the increasing energy demands by synthesizing a new product called gasohol. However, both the single-cell protein and gasohol projects failed to progress due to varying issues including public resistance, a changing economic scene, and shifts in political power.

Yet the formation of a new field, genetic engineering, would soon bring biotechnology to the forefront of science in society, and the intimate relationship between the scientific community, the public, and the government would ensue. These debates gained exposure in 1975 at the Asilomar Conference, where Joshua Lederberg was the most outspoken supporter for this emerging field in biotechnology. By as early as 1978, with the synthesis of human insulin, Lederberg's claims would prove valid, and the biotechnology industry grew rapidly. Each new scientific advance became a media event designed to capture public support, and by the 1980s, biotechnology grew into a promising real industry. In 1988, only five proteins from genetically engineered cells had been approved as drugs by the FDA, but this number would skyrocket to over 125 by the end of the 1990s.

The field of genetic engineering remains a heated topic of discussion in today's society with the advent of gene therapy, stem cell research, cloning, and genetically-modified food. While it seems only natural nowadays to link pharmaceutical drugs as solutions to health and societal problems, this relationship of biotechnology serving social needs began centuries ago.

ORIGINS OF BIOTECHNOLOGY

Biotechnology arose from the field of zymotechnology, which began as a search for a better understanding of industrial fermentation, particularly beer. Beer was an important industrial, and not just social, commodity. In late 19th century Germany, brewing contributed as much to the gross national product as steel, and taxes on alcohol proved to be significant sources of revenue to the government. In the 1860s, institutes and remunerative consultancies were dedicated to the technology of brewing. The most famous was the private Carlsberg Institute, founded in 1875, which employed Emil Christian Hansen, who pioneered the pure yeast process for the reliable production of consistent beer. Less well known were private consultancies that advised the brewing industry. One of these, the Zymotechnic Institute, was established in Chicago by the German-born chemist John Ewald Siebel.

The heyday and expansion of zymotechnology came in World War I in response to industrial needs to support the war. Max Delbruck grew yeast on an immense scale during the war to meet 60 percent of Germany's animal feed needs. Compounds of another fermentation product, lactic acid, made up for a lack of hydraulic fluid, glycerol. On the Allied side the Russian chemist Chaim Weizmann used starch to eliminate Britain's shortage of acetone, a key raw material in explosives, by fermenting maize to acetone. The industrial potential of fermentation was outgrowing its traditional home in brewing, and "zymotechnology" soon gave way to "biotechnology."

With food shortages spreading and resources fading, some dreamed of a new industrial solution. Karl Ereky coined the word "biotechnology" in Hungary during 1917 to describe a technology based on converting raw materials into a more useful product. He built a slaughterhouse for a thousand pigs and also a fattening farm with space for 50,000 pigs, raising over 100,000 pigs a year. The enterprise was enormous, becoming one of the largest and most profitable meat and fat operations in the world. In a book entitled Biotechnologie, Ereky further developed a theme that would be reiterated through the 20th century: biotechnology could provide solutions to societal crises, such as food and energy shortages. For Ereky, the term "biotechnologie" indicated the process by which raw materials could be biologically upgraded into socially useful products.

This catchword spread quickly after the First World War, as "biotechnology" entered German dictionaries and was taken up abroad by business-hungry private consultancies as far away as the United States. In Chicago, for example, the coming of prohibition at the end of World War I encouraged biological industries to create opportunities for new fermentation products, in particular a market for nonalcoholic drinks. Emil Siebel, the son of the founder of the Zymotechnic Institute, broke away from his father's company to establish his own called the "Bureau of Biotechnology," which specifically offered expertise in fermented nonalcoholic drinks. The belief that the needs of an industrial society could be met by fermenting agricultural waste was an important ingredient of the "chemurgic movement." Fermentation-based processes generated products of ever-growing utility. In the 1940s, penicillin was the most dramatic. While it was discovered in England, it was produced industrially in the U.S. using a deep fermentation process originally developed in Peoria, Illinois. The enormous profits and the public expectations penicillin engendered caused a radical shift in the standing of the pharmaceutical industry. Doctors used the phrase "miracle drug," and the historian of its wartime use, David Adams, has suggested that to the public penicillin represented the perfect health that went together with the car and the dream house of wartime American advertising. In the 1950s, steroids were synthesized using fermentation technology. In particular, cortisone promised the same revolutionary ability to change medicine as penicillin had.

SINGLE-CELL PROTEIN AND GASOHOL PROJECTS

Even greater expectations of biotechnology were raised during the 1960s by a process that grew single-cell protein. When the so-called protein gap threatened world hunger, producing food locally by growing it from waste seemed to offer a solution. It was the possibilities of growing microorganisms on oil that captured the imagination of scientists, policy makers, and commerce.

Major companies such as British Petroleum (BP) staked their futures on it. In 1962, BP built a pilot plant at Cap de Lavera in Southern France to publicize its product, Toprina. As there was no well-accepted term to describe the new foods, in 1966 the term "single-cell protein" (SCP) was coined at MIT to provide an acceptable and exciting new title, avoiding the unpleasant connotations of microbial or bacterial.

By the late 1970s, however, the cultural climate had completely changed, as the growth in SCP interest had taken place against a shifting economic and cultural scene (136). First, the price of oil rose catastrophically in 1974, so that its cost per barrel was five times greater than it had been two years earlier. Second, despite continuing hunger around the world, anticipated demand also began to shift from humans to animals. The program had begun with the vision of growing food for Third World people, yet the product was instead launched as an animal food for the developed world. The rapidly rising demand for animal feed made that market appear economically more attractive. The ultimate downfall of the SCP project, however, came from public resistance. This was particularly vocal in Japan, where production came closest to fruition.

For all their enthusiasm for innovation and traditional interest in microbiologically produced foods, the Japanese were the first to ban the production of single-cell proteins. The Japanese ultimately were unable to separate the idea of their new "natural" foods from the far from natural connotation of oil. These arguments were made against a background of suspicion of heavy industry in which anxiety over minute traces of petroleum was expressed. Thus, public resistance to an unnatural product led to the end of the SCP project as an attempt to solve world hunger.

In the late 1970s, biotechnology offered another possible solution to a societal crisis. The escalation in the price of oil in 1974 increased the cost of the Western world's energy tenfold. In response, the U.S. government promoted the production of gasohol, gasoline with 10 percent alcohol added, as an answer to the energy crisis. In 1979, when the Soviet Union sent troops to Afghanistan, the Carter administration cut off its supplies to agricultural produce in retaliation, creating a surplus of agriculture in the U.S. As a result, fermenting the agricultural surpluses to synthesize fuel seemed to be an economical solution to the shortage of oil threatened by the Iran-Iraq war. Before the new direction could be taken, however, the political wind changed again: the Reagan administration came to power in January 1981 and, with the declining oil prices of the 1980s, ended support for the gasohol industry before it was born.

Biotechnology seemed to be the solution for major social problems, including world hunger and energy crises. In the 1960s, radical measures would be needed to meet world starvation, and biotechnology seemed to provide an answer. However, the solutions proved to be too expensive and socially unacceptable, and solving world hunger through SCP food was dismissed. In the 1970s, the food crisis was succeeded by the energy crisis, and here too, biotechnology seemed to provide an answer. But once again, costs proved prohibitive as oil prices slumped in the 1980s. Thus, in practice, the implications of biotechnology were not fully realized in these situations. But this would soon change with the rise of genetic engineering.

GENETIC ENGINEERING

The origins of biotechnology culminated with the birth of genetic engineering. There were two key events that have come to be seen as scientific breakthroughs beginning the era that would unite genetics with biotechnology. One was the 1953 discovery of the structure of DNA, by Watson and Crick, and the other was the 1973 discovery by Cohen and Boyer of a recombinant DNA technique by which a section of DNA was cut from the plasmid of an E. coli bacterium and transferred into the DNA of another. This approach could, in principle, enable bacteria to adopt the genes and produce proteins of other organisms, including humans. Popularly referred to as "genetic engineering," it came to be defined as the basis of new biotechnology.

Genetic engineering proved to be a topic that thrust biotechnology into the public scene, and the interaction between scientists, politicians, and the public defined the work that was accomplished in this area. Technical developments during this time were revolutionary and at times frightening. In December 1967, the first heart transplant by Christian Barnard reminded the public that the physical identity of a person was becoming increasingly problematic. While poetic imagination had always seen the heart at the center of the soul, now there was the prospect of individuals being defined by other people's hearts. During the same month, Arthur Kornberg announced that he had managed to biochemically replicate a viral gene. "Life had been synthesized," said the head of the National Institutes of Health. Genetic engineering was now on the scientific agenda, as it was becoming possible to identify genetic characteristics with diseases such as beta thalassemia and sickle-cell anemia. Responses to scientific achievements were colored by cultural skepticism. Scientists and their expertise were looked upon with suspicion. In 1968, an immensely popular work, The Biological Time Bomb, was written by the British journalist Gordon Rattray Taylor. The author's preface saw Kornberg's discovery of replicating a viral gene as a route to lethal doomsday bugs. The publisher's blurb for the book warned that within ten years, "You may marry a semi-artificial man or woman..choose your children's sex..tune out pain..change your memories..and live to be 150 if the scientific revolution doesn't destroy us first." The book ended with a chapter called "The Future-If Any." While it is rare for current science to be represented in the movies, in this period of "Star Trek", science fiction and science fact seemed to be converging. "Cloning" became a popular word in the media. Woody Allen satirized the cloning of a person from a nose in his 1973 movie Sleeper, and cloning Hitler from surviving cells was the theme of the 1976 novel by Ira Levin, The Boys from Brazil.

In response to these public concerns, scientists, industry, and governments increasingly linked the power of recombinant DNA to the immensely practical functions that biotechnology promised. One of the key scientific figures that attempted to highlight the promising aspects of genetic engineering was Joshua Lederberg, a Stanford professor and Nobel laureate. While in the 1960s "genetic engineering" described eugenics and work involving the manipulation of the human genome, Lederberg stressed research that would involve microbes instead. Lederberg emphasized the importance of focusing on curing living people. Lederberg's 1963 paper, "Biological Future of Man" suggested that, while molecular biology might one day make it possible to change the human genotype, "what we have overlooked is euphenics, the engineering of human development." Lederberg constructed the word "euphenics" to emphasize changing the phenotype after conception rather than the genotype which would affect future generations.

With the discovery of recombinant DNA by Cohen and Boyer in 1973, the idea that genetic engineering would have major human and societal consequences was born. In July 1974, a group of eminent molecular biologists headed by Paul Berg wrote to Science suggesting that the consequences of this work were so potentially destructive that there should be a pause until its implications had been thought through. This suggestion was explored at a meeting in February 1975 at California's Monterey Peninsula, forever immortalized by the location, Asilomar. Its historic outcome was an unprecedented call for a halt in research until it could be regulated in such a way that the public need not be anxious, and it led to a 16-month moratorium until NIH guidelines were established.

Lederberg was the leading exception in emphasizing, as he had for years, the potential benefits. At Asilomar, in an atmosphere favoring control and regulation, he circulated a paper countering the pessimism and fears of misuses with the benefits conferred by successful use. He described "an early chance for a technology of untold importance for diagnostic and therapeutic medicine: the ready production of an unlimited variety of human proteins. Analogous applications may be foreseen in fermentation process for cheaply manufacturing essential nutrients, and in the improvement of microbes for the production of antibiotics and of special industrial chemicals." In June 1976, the 16-month moratorium on research expired with the Director's Advisory Committee (DAC) publication of the NIH guidelines of good practice. They defined the risks of certain kinds of experiments and the appropriate physical conditions for their pursuit, as well as a list of things too dangerous to perform at all. Moreover, modified organisms were not to be tested outside the confines of a laboratory or allowed into the environment.

Atypical as Lederberg was at Asilomar, his optimistic vision of genetic engineering would soon lead to the development of the biotechnology industry. Over the next two years, as public concern over the dangers of recombinant DNA research grew, so too did interest in its technical and practical applications. Curing genetic diseases remained in the realms of science fiction, but it appeared that producing human proteins would be good business. Insulin, one of the smaller and best understood proteins, had been used in treating diabetes for half a century. It had been extracted from animals in a chemically slightly different form from the human product. Yet, if one could produce human insulin, one could meet an existing enormous demand with an apparently superior product whose approval would be relatively easy. In the period 1975 to 1977, insulin represented the aspirations for new products that could be made with the new biotechnology. Microbial production of human insulin was finally announced in September 1978 and was commercially produced by a startup company, Genentech.

The radical shift in the connotation of "genetic engineering" from an emphasis on the inherited characteristics of people to the commercial production of proteins and therapeutic drugs was nurtured by Joshua Lederberg. His broad concerns since the 1960s had been stimulated by enthusiasm for science and its potential medical benefits. Countering calls for strict regulation, he expressed a vision of potential utility. Against a belief that new techniques would entail unmentionable and uncontrollable consequences for humanity and the environment, a growing consensus on the economic value of recombinant DNA emerged.

BIOTECHNOLOGY AND INDUSTRY

With ancestral roots in industrial microbiology that date back centuries, the new biotechnology industry grew rapidly beginning in the mid-1970s. Each new scientific advance became a media event designed to capture investment confidence and public support. Although market expectations and social benefits of new products were frequently overstated, many people were prepared to see genetic engineering as the next great advance in technological progress. By the 1980s, biotechnology characterized a nascent real industry, providing titles for emerging trade organizations such as the Industrial Biotechnology Association.

The main foci of attention after insulin were the potential profit makers in the pharmaceutical industry: human growth hormone and what promised to be a miraculous cure for viral diseases, interferon. Cancer was a central target in the 1970s because increasingly the disease was linked to viruses. By 1980, a new company, Biogen, had produced interferon through recombinant DNA. The emergence of interferon and the possibility of curing cancer raised money in the community for research and increased the enthusiasm of an otherwise uncertain and tentative society.

Moreover, to the 1970s plight of cancer was added AIDS in the 1980s, offering an enormous potential market for a successful therapy, and more immediately, a market for diagnostic tests based on monoclonal antibodies. By 1988, only five proteins from genetically engineered cells had been approved as drugs by the FDA: insulin, human growth hormone, hepatitis B vaccine, alpha-interferon, and tissue plasminogen activator (tPA), for lysis of blood clots. By the end of the 1990s, however, 125 more genetically engineered drugs would be approved.

Genetic engineering also reached the agricultural front as well. There was tremendous progress since the market introduction of the genetically engineered Flavr Savr tomato in 1994. Ernst and Young reported that in 1998, 30% of the U.S. soybean crop was expected to be from genetically engineered seeds. In 1998, about 30% of the US cotton and corn crops were also expected to be products of genetic engineering.

Genetic engineering in biotechnology stimulated hopes for both therapeutic proteins, drugs and biological organisms themselves, such as seeds, pesticides, engineered yeasts, and modified human cells for treating genetic diseases. From the perspective of its commercial promoters, scientific breakthroughs, industrial commitment, and official support were finally coming together, and biotechnology became a normal part of business. No longer were the proponents for the economic and technological significance of biotechnology the iconoclasts. Their message had finally become accepted and incorporated into the policies of governments and industry.

7

THE SCIENTIFIC STUDY OF ANIMAL BEHAVIOUR

Ethology is the scientific study of animal behavior considered as a branch of zoology. A scientist who practices ethology is called an ethologist.

The desire to understand the animal world has made ethology a rapidly growing field, and even since the turn of the 21st century, many prior understandings related to diverse fields such as animal communication, personal symbolic name use, animal emotions, animal culture and learning, and even sexual conduct, long thought to be well understood, have been revolutionized, as have new fields such as neuroethology.

ETYMOLOGY

The term "ethology" is derived from the Greek word "ethos", meaning "custom." Other words derived from the Greek word "ethos" include "ethics" and "ethical." The term was first popularized in English by the American myrmecologist William Morton Wheeler in 1902. An earlier, slightly different sense of the term was proposed by John Stuart Mill in his 1843 System of Logic.

He recommended the development of a new science, "ethology," whose purpose would be the explanation of individual and national differences in character, on the basis of associationistic psychology. This use of the word for this purpose was never adopted.

DIFFERENCES AND SIMILARITIES

Comparative psychology also studies animal behavior, but, as opposed to ethology, construes its study as a branch of psychology rather than as one of biology. Thus, where comparative psychology sees the study of animal behavior in the context of what is known about human psychology, ethology sees the study of animal behavior in the context of what is known about animal anatomy and physiology. Furthermore, early comparative psychologists concentrated on the study of learning and tended to look at behavior in artificial situations, whereas early ethologists concentrated on behavior in natural situations, tending to describe it as instinctive. The two approaches are complementary rather than competitive, but they do lead to different perspectives and sometimes to conflicts of opinion about matters of substance.

In addition, for most of the twentieth century, comparative psychology developed most strongly in North America, while ethology was stronger in Europe, and this led to different emphases as well as somewhat differing philosophical underpinnings in the two disciplines. A practical difference is that early comparative psychologists concentrated on gaining extensive knowledge of the behavior of very few species, while ethologists were more interested in gaining knowledge of behavior in a wide range of species in order to be able to make principled comparisons across taxonomic groups. Ethologists have made much more use of a truly comparative method than comparative psychologists ever have. Despite the historical divergence, most ethologists (as opposed to behavioural ecologists), at least in North America, teach in psychology departments.

DARWINISM AND THE BEGINNINGS OF ETHOLOGY

Because ethology is understood as a branch of biology, ethologists have been particularly concerned with the evolution of behavior and the understanding of behavior in terms of the theory of natural selection. In one sense, the first modern ethologist was Charles Darwin, whose book, The expression of the emotions in animals and men, has influenced many ethologists. He pursued his interest in behavior by encouraging his protégé George

Romanes, who investigated animal learning and intelligence using an anthropomorphic method, anecdotal cognitivism, that did not gain scientific support. Other early ethologists, such as Oskar Heinroth and Julian Huxley, instead concentrated on behaviors that can be called instinctive, or natural, in that they occur in all members of a species under specified circumstances. Their first step in studying the behavior of a new species was to construct an ethogram (a description of the main types of natural behavior with their frequencies of occurrence). This approach provided an objective, cumulative base of data about behavior, which subsequent researchers could check and build on.

The Fixed Action Pattern and Animal Communication

An important step, associated with the name of Konrad Lorenz though probably due more to his teacher, Oskar Heinroth, was the identification of fixed action patterns (FAPs). Lorenz popularized FAPs as instinctive responses that would occur reliably in the presence of identifiable stimuli (called sign stimuli or releasing stimuli). These FAPs could then be compared across species, and the similarities and differences between behavior could be easily compared with the similarities and differences in morphology. An important and much quoted study of the Anatidae (ducks and geese) by Heinroth used this technique. The ethologists noted that the stimuli that released FAPs were commonly features of the appearance or behavior of other members of their own species, and they were able to show how important forms of animal communication could be mediated by a few simple FAPs. The most sophisticated investigation of this kind was the study by Karl von Frisch of the so-called "dance language" underlying bee communication. Lorenz developed an interesting theory of the evolution of animal communication based on his observations of the nature of fixed action patterns and the circumstances in which animals emit them.

Imprinting

A second important finding of Lorenz concerned the early learning of young nidifugous birds, a process he called imprinting.

Lorenz observed that the young of birds such as geese and chickens spontaneously followed their mothers from almost the first day after they were hatched, and he discovered that this response could be imitated by an arbitrary stimulus if the eggs were incubated artificially and the stimulus was presented during a critical period (a less temporally constrained period is called a sensitive period) that continued for a few days after hatching.

Tinbergen's Four Questions for Ethologists

Lorenz's collaborator, Niko Tinbergen, argued that ethology always needed to pay attention to four kinds of explanation in any instance of behavior:

- *Function:* How does the behavior impact on the animal's chances of survival and reproduction?
- *Causation:* What are the stimuli that elicit the response, and how has it been modified by recent learning?
- *Development:* How does the behavior change with age, and what early experiences are necessary for the behavior to be shown?
- *Evolutionary History:* How does the behavior compare with similar behavior in related species, and how might it have arisen through the process of phylogeny?

THE FLOWERING OF ETHOLOGY

Through the work of Lorenz and Tinbergen, ethology developed strongly in continental Europe in the years before World War II. After the war, Tinbergen moved to the University of Oxford, and ethology became stronger in the UK, with the additional influence of William Thorpe, Robert Hinde, and Patrick Bateson at the Sub-department of Animal Behaviour of the University of Cambridge, located in the village of Madingley. In this period, too, ethology began to develop strongly in North America.

Lorenz, Tinbergen, and von Frisch were jointly awarded the Nobel Prize in 1973 for their work in developing ethology.

Social Ethology and Recent Developments

In 1970, the English ethologist John H. Crook published an important paper in which he distinguished comparative ethology from social ethology, and argued that much of the ethology that had existed so far was really comparative ethology--looking at animals as individuals--whereas in the future ethologists would need to concentrate on the behaviour of social groups of animals and the social structure within them.

Indeed, E. O. Wilson's book Sociobiology appeared in 1975, and since that time the study of behavior has been much more concerned with social aspects. It has also been driven by the stronger, but more sophisticated, Darwinism associated with Wilson and Richard Dawkins. The related development of behavioral ecology has also helped transform ethology. Furthermore, a substantial rapprochement with comparative psychology has occurred, so the modern scientific study of behavior offers a more or less seamless spectrum of approaches-from animal cognition to more traditional comparative psychology, ethology, sociobiology and behavioral ecology. Sociobiology has more recently developed in to evolutionary psychology.

ANTHROZOOLOGY

Anthrozoology is the study of human-animal interaction ("animal" referring to all non-human animals), also described as the science focusing on all aspects of the human-animal bond and a bridge between the natural and social sciences.

Anthrozoology is a modern multi-disciplinary and burgeoning field that overlaps with a number of other disciplines, including anthropology, ethology, medicine, psychology, veterinary medicine and zoology. A major focus of anthrozoologic research is the quantifying of the positive effects of human-animal relationships on either party and the study of the reality of their interactions.

"In reality, there is little homogeneity in anthrozoological research, certainly as far as disciplinary approach, methodology and theory is concerned. The academic backgrounds of anthrozoological researchers can be extremely varied. There are, for example, social scientists, psychologists, zoologists, ethologists, historians, philosophers, veterinarians and even physicians studying the human-animal relationship."

It adds that anthrozoology may veer towards ethics, behavioral research, welfare and promotion, and many other areas within the human-animal interface.

Example Areas of Study

- The interaction and enhancement within service animal interactions.
- Affective (emotional) or relational bonds between humans and animals (which may range from companion animals through to zoophilia)
- Human perceptions and beliefs in respect of other animals.
- How some animals fit into human societies.
- How these vary between cultures, and change over times.
- The study of animal domestication: how and why domestic animals evolved from wild species (paleoanthrozoology).

Examples of papers published by ISAZ as at 2005 include:

- Compatibility Stereotypes of People and Pets
- Behavior of Children with Learning Disabilities Interacting with a Therapy Dog
- Realistic Representations of Companion Animals in Comic Art in the USA
- Advertising Animal Protection
- Speciesism, Anthropocentrism, and Non-Western Cultures
- Whale Shark Tourism in Ningaloo Marine Park, Australia
- Judgments of Cruelty to Animals: Sex Differences and Effect of Awareness of Suffering
- Motivations and Characteristics of Volunteer Flying-Fox Rehabilitators in Australia
- Pets' Roles in Parents' Bereavement
- Recent Shifts Regarding the Relationship between People and Other Animals in Japanese Culture

- How Depressive Moods Affect the Behavior of Singly Living Persons Towards Their Cats

LIFE

Life can be a property of matter. Matter that is alive, or life forms, are organisms of a vast variety. Properties common to the known organisms found on Earth-plants, animals, fungi, protists, archaea and bacteria-are that they are carbon-and-water-based, are cellular with complex organization, undergo metabolism, possess a capacity to grow, respond to stimuli, reproduce and, through natural selection, adapt to their environment in succeeding generations.

An entity with the above properties is considered to be an organism. However, not every definition of life considers all of these properties to be essential. For example, the capacity for descent with modification is often taken as the only essential property of life. This definition notably includes viruses, which do not qualify under narrower definitions as they are acellular and do not metabolise. Broader definitions of life may also include theoretical non-carbon-based life and other alternative biology.

Although it cannot be pinpointed exactly, evidence suggests that life on Earth has existed for about 3.7 billion years .

DEFINITION BY OPINION

A Conventional Definition

Although there is no universal agreement on the definition of life, scientists generally accept that the biological manifestation of life exhibits the following phenomena:

1. *Homeostasis:* Regulation of the internal environment to maintain a constant state; for example, sweating to reduce temperature.
2. *Organization:* Being composed of one or more cells, which are the basic units of life.
3. *Metabolism:* Consumption of energy by converting nonliving material into cellular components (anabolism)

and decomposing organic matter (catabolism). Living things require energy to maintain internal organization (homeostasis) and to produce the other phenomena associated with life.

4. *Growth:* Maintenance of a higher rate of synthesis than catalysis. A growing organism increases in size in all of its parts, rather than simply accumulating matter. The particular species begins to multiply and expand as the evolution continues to flourish.
5. *Adaptation:* The ability to change over a period of time in response to the environment. This ability is fundamental to the process of evolution and is determined by the organism's heredity as well as the composition of metabolized substances, and external factors present.
6. *Response to Stimuli:* A response can take many forms, from the contraction of a unicellular organism when touched to complex reactions involving all the senses of higher animals. A response is often expressed by motion, for example, the leaves of a plant turning toward the sun or an animal chasing its prey.
7. *Reproduction:* The ability to produce new organisms. Reproduction can be the division of one cell to form two new cells. Usually the term is applied to the production of a new individual (either asexually, from a single parent organism, or sexually, from at least two differing parent organisms), although strictly speaking it also describes the production of new cells in the process of growth.

Exceptions to the Conventional Definition

It is important to note that life is a definition that applies primarily at the level of species, so even though many individuals of any given species do not reproduce, possibly because they belong to specialized sterile castes (such as ant workers), these are still considered forms of life. One could say that the property of life is inherited; hence, sterile hybrid species such as the mule are considered life although not themselves capable of reproduction.

It is also worth noting that non-reproducing individuals may still help the spread of their genes through such mechanisms as kin selection.

For similar reasons, viruses and aberrant prion proteins are often considered replicators rather than forms of life, a distinction warranted because they cannot reproduce without very specialized substrates such as host cells or proteins, respectively. However, most forms of life rely on foods produced by other species, or at least the specific chemistry of Earth's environment.

Some individuals contest such definitions of life on philosophical grounds, and offer the following as examples of life: viruses which reproduce; storms or flames which "burn"; certain computer software programs which are programmed to mutate and evolve; future software programs which may evince (even high-order) behavior; machines which can move; and some forms of proto-life consisting of metabolizing cells without the ability to reproduce. Still, most scientists would not call such phenomena expressive of life. Generally all seven characteristics are required for a population to be considered a life form.

Other Definitions

The systemic definition is that living things are self-organizing and autopoietic (self-producing). These objects are not to be confused with dissipative structures (e.g. fire).

Variations of this definition include Stuart Kauffman's definition of life as an autonomous agent or a multi-agent system capable of reproducing itself or themselves, and of completing at least one thermodynamic work cycle.

Another definition is: "Living things are systems that tend to respond to changes in their environment, and inside themselves, in such a way as to promote their own continuation."

Yet another definition: "Life is a self-organizing, cannibalistic system consisting of a population of replicators that are capable of mutation, around most of which homeostatic, metabolizing organisms evolve." This definition does not include flames, but does include worker ants, viruses and mules. Without 'most of', it does not include viruses. Self reproduction and energy consumption is only one means for a system to promote its own continuation. This explains why bees can be alive and yet commit suicide in defending their hive. In this case the whole colony works as such a living system.

Another attempt at defining the general features and properties of life: "type of organization of matter producing various interacting forms of variable complexity, whose main property is to replicate almost perfectly by using matter and energy available in their environment to which they may adapt."

In this definition "almost perfectly" relates to mutations happening during replication of organisms that may have adaptative benefits.

Contradictions in the Definition

Most of the definitions are focused on differencing the living material from the no-living one ,. Some new ideas were questioned in the years 1980-1990. It is still not clear whether exists a computer software that can be considered as a living one. Some computer software show certain resemblance with life. It is not difficult in algorithmic sense to create software which fulfills some base criteria (growing, reproduction, reaction to the changes in the environment, etc.).

Consider an artificial environment with certain properties (for example an system which is calling the processes with different input). If the process is able to solve the problem (respond with a correct output) in a given time, it can survive, otherwise it is deleted. If the process solves the problem before the other processes it gets higher part from the computer resources. This is a very simple model of an environment in which the life has been created. More sophisticated environments can be created. Although in most of the simulations the systems collapse relatively fast (for example non of the processes are able to adapt and solve the given problem) it has not yet been formally proved that such an environment can not exist permanently. Such an environment could met all the criteria of some simple life definitions.

If ever such a theorem will be formally proved, it will be necessary to reconsider the definitions or admit that the in some sense the computer world is a good approximation of the real one.

ORIGIN OF LIFE

There is no truly "standard" model for the origin of life, but most currently accepted scientific models build in one way or another on the following discoveries, which are listed roughly in order of postulated emergence:

1. Plausible pre-biotic conditions result in the creation of the basic small molecules of life. This was demonstrated in the Miller-Urey experiment.
2. Phospholipids spontaneously form lipid bilayers, the basic structure of a cell membrane.
3. Procedures for producing random RNA molecules can produce ribozymes, which are able to produce more of themselves under very specific conditions.

There are many different hypotheses regarding the path that might have been taken from simple organic molecules to protocells and metabolism. Many models fall into the "genes-first" category or the "metabolism-first" category, but a recent trend is the emergence of hybrid models that do not fit into either of these categories.

The Possibility of Extraterrestrial Life

Earth is the only planet in the universe known to harbour life. The Drake equation has been used to estimate the probability of life elsewhere, but scientists disagree on many of the values of variables in this equation (although strictly speaking Drake equation estimates the number of extraterrestrial civilizations in our galaxy with which we might come in contact-not probability of life elsewhere). Depending on those values, the equation may either suggest that life arises frequently or infrequently.

SCIENTIFIC CLASSIFICATION

Scientific classification or biological classification is a method by which biologists group and categorize species of organisms.

Scientific classification can also be called scientific taxonomy, but should be distinguished from folk taxonomy, which lacks scientific basis. Modern classification has its root in the work of Carolus Linnaeus, who grouped species according to shared physical characteristics. These groupings have since been revised to improve consistency with the Darwinian principle of common descent. Molecular systematics, which uses DNA sequences as data, has driven many recent revisions and is likely to continue to do so. Scientific classification belongs to the science of taxonomy or biological systematics.

Early Systems

The earliest known system of classifying forms of life comes from the Greek philosopher Aristotle, who classified all living organisms known at that time as either a plant or an animal. He further classified animals based on their means of transportation (air, land, or water).

In 1172 Ibn Rushd (Averroes), who was a judge (Qadi) in Seville, translated and abridged Aristotle's book de Anima (On the Soul) into Arabic. His original commentary is now lost, but its translation into Latin by Michael Scot survives.

An important advance was made by the Swiss professor, Conrad von Gesner (1516-1565). Gesner's work was a critical compilation of life known at the time.

The exploration of parts of the New World next brought to hand descriptions and specimens of many novel forms of animal life. In the latter part of the 16th century and the beginning of the 17th, careful study of animals commenced, which, directed first to familiar kinds, was gradually extended until it formed a sufficient body of knowledge to serve as an anatomical basis for classification. Advances in using this knowledge to classify living beings bear a debt to the research of medical anatomists, such as Fabricius (1537-1619), Petrus Severinus (1580-1656), William Harvey (1578-1657), and Edward Tyson (1649-1708). Advances in classification due to the work of entomologists and the first microscopists is due to the research of people like Marcello Malpighi (1628-1694), Jan

Swammerdam (1637-1680), and Robert Hooke (1635-1702). Lord Monboddo (1714-1799) was one of the early abstract thinkers whose works illustrate knowledge of species relationships and who foreshadowed the theory of evolution.

Early Methodists

Since late in the 15th century, a number of authors had become concerned with what they called methodus, or method. By method they meant an arrangement of minerals, plants, and animals according to the principles of logical division. The term methodists was coined by Carolus Linnaeus in his Bibliotheca Botanica to denote the authors who care about the principles of classification (in contrast to the mere collectors who are concerned primarily with the description of plants paying little or no attention to their arrangement into genera, etc).

Important early methodists were an Italian philosopher, physician, and botanist Andrea Caesalpino, an English naturalist John Ray, a German physician and botanist Augustus Quirinus Rivinus, and a French physician, botanist, and traveller Joseph Pitton de Tournefort.

Andrea Caesalpino (1519-1603) in his De plantis libri XVI (1583) proposed the first methodical arrangement of plants. On the basis of the structure of trunk and fructification he divided plants into fifteen "higher genera".

John Ray (1627-1705) was an English naturalist who published important works on plants, animals, and natural theology. The approach he took to the classification of plants in his Historia Plantarum was an important step towards modern taxonomy. Ray rejected the system of dichotomous division by which species were classified according to a pre-conceived, either/or type system, and instead classified plants according to similarities and differences that emerged from observation.

Both Caesalpino and Ray used traditional plant names and thus, the name of a plant did not reflect its taxonomic position (e.g. even though the apple and the peach belonged to different "higher genera" of John Ray's methodus, both retained their traditional names Malus and Malus Persica respectively). A further step was taken by Rivinus and Pitton de Tournefort who made genus a distinct rank within taxonomic hierarchy and introduced the practice of naming the plants according to their genera.

Augustus Quirinus Rivinus (1652-1723), in his classification of plants based on the characters of the flower, introduced the category of order (corresponding to the "higher" genera of John Ray and Andrea Caesalpino). He was the first to abolish the ancient division of plants into herbs and trees and insisted that the true method of division should be based on the parts of the fructification alone.

Rivinus extensively used dichotomous keys to define both orders and genera. His method of naming plant species resembled that of Joseph Pitton de Tournefort. The names of all plants belonging to the same genus should begin with the same word (generic name). In the genera containing more than one species the first species was named with generic name only, while the second, etc were named with a combination of the generic name and a modifier (differentia specifica).

Joseph Pitton de Tournefort (1656-1708) introduced an even more sophisticated hierarchy of class, section, genus, and species. He was the first to use consistently the uniformly composed species names which consisted of a generic name and a many-worded diagnostic phrase differentia specifica. Unlike Rivinus, he used differentiae with all species of polytipic genera.

Linnaeus

Two years after John Ray's death, Carolus Linnaeus (1707-1778) was born. His great work, the Systema Naturae, ran through twelve editions during his lifetime (1st ed. 1735). In this work, nature was divided into three kingdoms: mineral, vegetable and animal. Linnaeus used five ranks: class, order, genus, species, and variety.

He abandoned long descriptive names of classes and orders and two-word generic names (e.g. Bursa pastoris) still used by his immediate predecessors (Rivinus and Pitton de Tournefort) and replaced them with single-word names, provided genera with detailed diagnoses (characteres naturales), and reduced numerous varieties to their species, thus saving botany from the chaos of new forms produced by horticulturalists.

Linnaeus is best known for his introduction of the method still used to formulate the scientific name of every species. Before Linnaeus, long many-worded names (composed of a generic name and a differentia specifica) had been used, but as these names gave a description of the species, they were not fixed. In his Philosophia Botanica (1751) Linnaeus took every effort to improve the composition and reduce the length of the many-worded names by abolishing unnecessary rhetorics, introducing new descriptive terms and defining their meaning with an unprecedented precision.

In the late 1740s Linnaeus began to use a parallel system of naming species with nomina trivialia. Nomen triviale, a trivial name, was a single-or two-word epithet placed on the margin of the page next to the many-worded "scientific" name. The only rules Linnaeus applied to them was that the trivial names should be short, unique within a given genus, and that they should not be changed. Linnaeus consistently applied nomina trivialia to the species of plants in Species Plantarum (1st edn. 1753) and to the species of animals in the 10th edition of Systema Naturae (1758).

By consistently using these specific epithets, Linnaeus separated nomenclature from taxonomy. Even though the parallel use of nomina trivialia and many-worded descriptive names continued until late in the eighteenth century, it was gradually replaced by the practice of using shorter proper names combined of the generic name and the trivial name of the species. In the nineteenth century, this new practice was codified in the first Rules and Laws of Nomenclature, and the 1st edn. of Species Plantarum and the 10th edn. of Systema Naturae were chosen as starting points for the Botanical and Zoological Nomenclature respectively. This convention for naming species is referred to as binomial nomenclature. Today, nomenclature is regulated by Nomenclature Codes, which allows names divided into ranks; see rank (botany) and rank (zoology).

MODERN DEVELOPMENTS

Whereas Linnaeus classified for ease of identification, it is now generally accepted that classification should reflect the Darwinian principle of common descent. Since the 1960s a trend called cladistic taxonomy (or cladistics or cladism) has emerged, arranging taxa in an evolutionary tree. If a taxon includes all the descendants of some ancestral form, it is called monophyletic, as opposed to paraphyletic. Other groups are called polyphyletic. A new formal code of nomenclature, the PhyloCode, is currently under development, intended to deal with clades rather than taxa. It is unclear, should this be implemented, how the different codes will coexist. Domains are a relatively new grouping. The three-domain system was first invented in 1990, but not generally accepted until later. Now, the majority of biologists accept the domain system, but a large minority use the five-kingdom method. One main characteristic of the three-domain method is the separation of Archaea and Bacteria, previously grouped into the single kingdom Bacteria (sometimes Monera). A small minority of scientists add Archaea as a sixth kingdom but do not accept the domain method.

Authorities

The name of any taxon may be followed by the "authority" for the name, that is, the name of the author who first published a valid description of it. These names are frequently abbreviated: the abbreviation "L." is universally accepted for Linnaeus, and in botany there is a regulated list of standard abbreviations. The system for assigning authorities is slightly different in different branches of biology: see author citation (botany) and author citation (zoology). However, it is standard that if a name or placement has been changed since the original description, the first authority's name is placed in parentheses and the authority for the new name or placement may be placed after it (usually only in botany).

DOMAIN (BIOLOGY)

In biology, domain (also superregnum, superkingdom, or empire) is the top-level taxon of organisms in scientific classification, higher than a kingdom. Domain (or its synonyms) is the most inclusive of these biological groupings. The names are arbitrary, but the arrangement reflects the fundamental evolutionary differences in the genomes. There are several modern alternative domain classifications of life. Among them are:

- The two-empire system, with top-level groupings of Prokaryota (or Monera) and Eukaryota empires.
- The six-kingdom system with top-level groupings of Protista, Archaebacteria, Eubacteria, Fungi, Plantae, and Animalia.

and the most recent,

- The three-domain system of Carl Woese, introduced in 1990, with top-level groupings of Archaea, Bacteria, and Eukaryota domains.

As these groupings depend primarily on the analysis of genetic sequence data, additional proposed arrangements are to be expected.

KINGDOM (BIOLOGY)

In biology, a kingdom or regnum is the top-level, or nearly the top-level, taxon of organisms in scientific classification. (Sometimes domain or empire has been used as the topmost level.) The kingdoms are then divided into smaller groups called phyla (for animals) or divisions (for plants).

In his Systema Naturae, first published in 1735, Carolus Linnaeus distinguished two kingdoms of living things: Animalia for animals and Vegetabilia for plants (Linnaeus also treated minerals, placing them in a third kingdom, Mineralia).

Linnaeus divided each kingdom into classes, later grouped into phyla for animals and divisions for plants. When single-celled organisms were first discovered, they were split between the two kingdoms: mobile forms in the animal phylum Protozoa, and colored algae and bacteria in the plant division Thallophyta or Protophyta. However, a number of forms were hard to place, or were placed in different kingdoms by different authors: for example, the mobile alga Euglena and the amoeba-like slime moulds. As a result, Ernst Haeckel suggested creating a third kingdom Protista for them.

Two Empires, Four Kingdoms

The discovery that bacteria have a radically different cell structure from other organisms-the bacterial cell has one or two membranes that lie at or near its surface, whereas other organisms have a more complex structure with a nucleus and other organelles divided by intracellular membranes-led Chatton to propose a division of life into two empires: organisms with a nucleus in Eukaryota and organisms without in Prokaryota.

Chatton's proposal was not taken up immediately; a more typical system was that of Herbert Copeland, who gave the prokaryotes a separate kingdom, originally called Mychota but later referred to as Monera or Bacteria. Copeland's four-kingdom system placed all eukaryotes other than animals and plants in the kingdom Protista

It gradually became apparent how important the prokaryote/ eukaryote distinction is, and Stanier and van Niel popularized Chatton's two-empire system in the 1960s.

Five Kingdoms

Robert Whittaker recognized an additional kingdom for the Fungi. The resulting five-kingdom system, proposed in 1969, has become a popular standard and with some refinement is still used in many works, or forms the basis for newer multi-kingdom systems. It is based mainly on differences in nutrition: his Plantae were mostly multicellular autotrophs, his Animalia multicellular heterotrophs, and his Fungi multicellular saprotrophs. The remaining two kingdoms, Protista and Monera, included unicellular and simple cellular colonies.

Six Kingdoms

In the years around 1980 there was an emphasis on phylogeny and redefining the kingdoms to be monophyletic groups, groups made up of relatively closely related organisms. The Animalia,

Plantae, and Fungi were generally reduced to core groups of closely related forms, and the others thrown into the Protista. Based on rRNA studies Carl Woese divided the prokaryotes (Kingdom Monera) into two kingdoms, called Eubacteria and Archaebacteria. Carl Woese attempted to establish a Three Kingdom system in which Plants, Animals, Protista, and Fungi were lumped into one kingdom of all eukaryotes. The Eubacteria and Archaebacteria made up the other two kingdoms. The initial use of "six Kingdom systems" represents a blending of the classic Five Kingdom system and Woese's Three Kingdom system. Such six Kingdom systems have become standard in many works .

A variety of new eukaryotic kingdoms were also proposed, but most were quickly invalidated, ranked down to phyla or classes, or abandoned. The only one which is still in common use is the kingdom Chromista proposed by Cavalier-Smith, including organisms such as kelp, diatoms, and water moulds. Thus the eukaryotes are divided into three primarily heterotrophic groups, the Animalia, Fungi, and Protozoa, and two primarily photosynthetic groups, the Plantae (including red and green algae) and Chromista. However, it has not become widely used because of uncertainty over the monophyly of the latter two kingdoms.

Woese stresses genetic similarity over outward appearance, relying on comparisons of ribosomal RNA genes at the molecular level to sort out classification catagories. A plant does not look like an animal, but at the cellular level, both groups are eukaryotes, having similar subcellular organization, including cell nuclei, which the Eubacteria and Archaebacteria do not have. More importantly, plants, animals, fungi, and protists are more similar to each other in their genetic makeup at the molecular level, based on rRNA studies, than they are to either the Eubacteria or Archaebacteria. Woese also found that all of the eukaryotes, lumped together as one group, are more closely related, genetically, to the Archaebacteria than they are to the Eubacteria. This means that the Eubacteria and Archaebacteria are separate groups even when compared to the eukaryotes. So, Woese established the Three-domain system, clarifying that all the Eukaryotes are more closely genetically related compared to their genetic relationship to either the bacteria or the archaebacteria, without having to replace the "six kingdom systems" with a three kingdom system.

The Three Domain system is a "six kingdom system" that unites the eukaryotic kingdoms into the Eukarya Domain based on their relative genetic similarity when compared to the Bacteria Domain and the Archaea Domain. Woese also recognized that the Protista Kingdom is not a monophyletic group and might be further divided at the level of Kingdom. Others have divided the Protista Kingdom into the Protozoa and the Chromista, for instance.

ORGANISM

In biology and ecology, an organism (in Greek organon = instrument) is a living complex adaptive system of organs that influence each other in such a way that they function in some way as a stable whole.

The origin of life on Earth and the relationships between its major lineages are controversial. Two main grades may be distinguished, the prokaryotes and eukaryotes. The prokaryotes are generally considered to represent two separate domains, called the Bacteria and Archaea, which are not closer to one another than to the eukaryotes. The gap between prokaryotes and eukaryotes is widely considered a major missing link in evolutionary history. Two eukaryotic organelles, namely mitochondria and chloroplasts, are generally considered to be derived from endosymbiotic bacteria. Fungi, animals and plants are examples of species that are eukaryote. Organisms all go through a life cycle.

The phrase complex organism describes any organism with more than one cell.

SEMANTICS

The word "organism" may broadly be defined as an assembly of molecules that influence each other in such a way that they function as a more or less stable whole and have properties of life. However, many sources, lexical and scientific, add conditions that are problematic to defining the word. The Oxford English Dictionary defines an organism as "[an] individual animal, plant, or single-celled life form" This definition problematically excludes non-animal and plant multi-cellular life forms such as some fungi and protista. Less controversially, perhaps, it excludes viruses and theoretically-possible man-made non-organic life forms.

Chambers Online Reference provides a much broader definition: "any living structure, such as a plant, animal, fungus or bacterium, capable of growth and reproduction". The definition emphasises life; it allows for any life form, organic or otherwise, to be considered an organism. This does encompass all cellular life, as well as possible synthetic life. This definition does lack anything approximating to the word "individual" which would exclude viruses.

The word "organism" usually describes an independent collections of systems (for example circulatory, digestive, or reproductive) themselves collections of organs; these are, in turn, collections of tissues, which are themselves made of cells. The concept of an organism can be challenged on grounds that organisms themselves are never truly independent of an ecosystem; groups or populations of organisms function in an ecosystem in a manner not unlike the function of multicellular tissues in an organism; when organisms enter into strict symbiosis, they are not independent in any sense that could not also be conferred upon an organ or a tissue. Symbiotic plant and algae relationships do consist of radically different DNA structures between contrasting groups of tissues, sufficient to recognize their reproductive independence.

However, in a similar way, an organ within an "organism" (say, a stomach) can have an independent and complex interdependent relationship to separate whole organisms, or groups of organisms (a population of viruses, or bacteria), without which the organ's stable function would transform or cease. Other organs within that system (say, the ribcage) might be affected only indirectly by such an arrangement, much the same way species' affect one another indirectly in an ecosystem. Thus, the boundaries of the organism are nearly always disputable, and all living matter exists within larger heterarchical systems of life, made of wide varieties of transient living and dead tissues, and functioning in complex and dynamic relationships to one another.

Viruses

Viruses are not typically considered to be organisms because they are not capable of "independent" reproduction or metabolism. This controversy is problematic, though, since some parasites and endosymbionts are also incapable of independent life. Although viruses have enzymes and molecules characteristic of living organisms, they are incapable of reproducing outside a host cell and most of their metabolic processes require a host and its 'genetic machinery.'

Superorganism

A superorganism is an organism consisting of many organisms. This is usually meant to be a social unit of eusocial animals, where division of labour is highly specialised and where individuals are not able to survive by themselves for extended periods of time. Ants are the most well known example of such a superorganism. Thermoregulation, a feature usually exhibited by individual organisms, does not occur in individuals or small groups of honeybees of the species Apis mellifera. When these bees pack together in clusters of between 5000 and 40000, the colony can thermoregulate. James Lovelock, with his "Gaia Theory" has paralleled the work of Vladimir Vernadsky, who suggested the whole of the biosphere in some respects can be considered as a superorganism. The concept of superorganism is under dispute, as many biologists maintain that in order for a social unit to be considered an organism by itself, the individuals should be in permanent physical connection to each other, and its evolution should be governed by selection to the whole society instead of individuals. While it's generally accepted that the society of eusocial animals is a unit of natural selection to at least some extent, most evolutionists claim that the individuals are still the primary units of selection.

The question remains "What is to be considered the individual?". Darwinians like Richard Dawkins suggest that the individual selected is the "Selfish Gene". Others believe it is the whole genome of an organism. E.O. Wilson has shown that with ant-colonies and other social insects it is the breeding entity of the colony that is selected, and not its individual members. This could apply to the bacterial members of a stromatolite, which, because of genetic sharing, in some way comprise a single gene pool. Gaian theorists like Lynn Margulis would argue this applies equally to the symbiogenesis of the bacterial underpinnings of the whole of the Earth.

It would appear, from computer simulations like Daisyworld that biological selection occurs at multiple levels simultaneously.

It is also argued that humans are actually a superorganism that includes microorganisms such as bacteria. It is estimated that "the human intestinal microbiota is composed of 1013 to 1014 microorganisms whose collective genome ("microbiome") contains at least 100 times as many genes as our own[...] Our microbiome has significantly enriched metabolism of glycans, amino acids, and xenobiotics; methanogenesis; and 2-methyl-D-erythritol 4-phosphate pathway-mediated biosynthesis of vitamins and isoprenoids. Thus, humans are superorganisms whose metabolism represents an amalgamation of microbial and human attributes."

Organizational Terminology

All organisms are classified by the science of alpha taxonomy into either taxa or clades.

Taxa are ranked groups of organisms which run from the general (domain) to the specific (species). A broad scheme of ranks in hierarchical order is:

- Domain
- Kingdom
- Phylum
- Class
- Order
- Family

- Genus
- Species

To give an example, Homo sapiens is the Latin binomial equating to modern humans. All members of the species sapiens are, at least in theory, genetically able to interbreed. Several species may belong to a genus, but the members of different species within a genus are unable to interbreed to produce fertile offspring. Homo, however, only has one surviving species (sapiens); Homo erectus, Homo neanderthalensis, &c. having become extinct thousands of years ago. Several genera belong to the same family and so on up the hierarchy. Eventually, the relevant kingdom (Animalia, in the case of humans) is placed into one of the three domains depending upon certain genetic and structural characteristics.

All living organisms known to science are given classification by this system such that the species within a particular family are more closely related and genetically similar than the species within a particular phylum.

Chemistry

Organisms are complex chemical reactions, organized in ways that promote reproduction and some measure of sustainability or survival. The molecular phenomena of chemistry are fundamental in understanding organisms, but it is a philosophical error (reductionism) to reduce organismal biology to mere chemistry. It is generally the phenomena of entire organisms that determine their fitness to an environment and therefore the survivability of their DNA based genes.

Organisms clearly owe their origin, metabolism, and many other internal functions to the phenomena at the level of chemistry, especially the chemistry of large organic molecules. Organisms are complex systems of chemical compounds which, through interaction with each other and the environment, play a wide variety of roles.

Organisms are semi-closed chemical systems. Although they are individual units of life (as the definition requires) they are not closed to the environment around them. To operate they constantly take in and release energy. Autotrophs produce usable energy (in the form of organic compounds) using light from the sun or inorganic compounds while heterotrophs take in organic compounds from the environment.

The primary chemical element in these compounds is carbon. The physical properties of this element such as its great affinity for bonding with other small atoms, including other carbon atoms, and its small size makes it capable of forming multiple bonds, make it ideal as the basis of organic life. It is able to form small compounds containing three atoms (such as carbon dioxide) as well as large chains of many thousands of atoms which are able to store data (nucleic acids), hold cells together and transmit information (protein).

Some branches of biology, especially ecology, do not gain significant benefit from reduction to chemical reactions.

Macromolecules

The compounds which make up organisms may be divided into macromolecules and other, smaller molecules. The four groups of macromolecule are nucleic acids, proteins, carbohydrates and lipids. Nucleic acids (specifically deoxyribonucleic acid, or DNA) store genetic data as a sequence of nucleotides. The particular sequence of the four different types of nucleotides (adenine, cytosine, guanine, and thymine) dictate the many characteristics which constitute the organism. The sequence is divided up into codons, each of which is a particular sequence of three nucleotides and corresponds to a particular amino acid. Thus a a sequence of DNA codes for a particular protein which, due to the chemical properties of the amino acids of which it is made, folds in a particular manner and so performs a particular function.

The following functions of protein have been recognized:

1. Enzymes, which catalyze all of the reactions of metabolism;
2. Structural proteins, such as tubulin, or collagen;
3. Regulatory proteins, such as transcription factors or cyclins that regulate the cell cycle;

4. Signalling molecules or their receptors such as some hormones and their receptors;
5. Defensive proteins, which can include everything from antibodies of the immune system, to toxins (e.g., dendrotoxins of snakes), to proteins that include unusual amino acids like canavanine.

Lipids make up the membrane of cells which constitutes a barrier, containing everything within the cell and preventing compounds from freely passing into, and out of, the cell. In some multi-cellular organisms they serve to store energy and mediate communication between cells. Carbohydrates also store and transport energy in some organisms, but are more easily broken down than lipids.

Structure

All organisms consist of monomeric units called cells; some contain a single cell (unicellular) and others contain many units (multicellular). Multicellular organisms are able to specialise cells to perform specific functions, a group of such cells is tissue the four basic types of which are epithelium, nervous tissue, muscle tissue and connective tissue. Several types of tissue work together in the form of an organ to produce a particular function (such as the pumping of the blood by the heart, or as a barrier to the environment as the skin). This pattern continues to a higher level with several organs functioning as an organ system to allow for reproduction, digestion, &c. Many multicelled organisms comprise of several organ systems which coordinate to allow for life.

The Cell

The cell theory, first developed in 1839 by Schleiden and Schwann, states that all organisms are composed of one or more cells; all cells come from preexisting cells; all vital functions of an organism occur within cells, and cells contain the hereditary information necessary for regulating cell functions and for transmitting information to the next generation of cells. There are two types of cells, eukaryotic and prokaryotic. Prokaryotic cells are usually singletons, while eukaryotic cells are usually found in multi-cellular organisms. Prokaryotic cells lack a nuclear membrane so DNA is unbound within the cell, eukaryotic cells have nuclear membranes. All cells, whether prokaryotic or eukaryotic, have a membrane, which envelopes the cell, separates its interior from its environment, regulates what moves in and out, and maintains the electric potential of the cell. Inside the membrane, a salty cytoplasm takes up most of the cell volume. All cells possess DNA, the hereditary material of genes, and RNA, containing the information necessary to build various proteins such as enzymes, the cell's primary machinery. There are also other kinds of biomolecules in cells. All cells share several abilities:

- Reproduction by cell division (binary fission, mitosis or meiosis).
- Use of enzymes and other proteins coded for by DNA genes and made via messenger RNA intermediates and ribosomes.
- Metabolism, including taking in raw materials, building cell components, converting energy, molecules and releasing by-products. The functioning of a cell depends upon its ability to extract and use chemical energy stored in organic molecules. This energy is derived from metabolic pathways.
- Response to external and internal stimuli such as changes in temperature, pH or nutrient levels.
- Cell contents are contained within a cell surface membrane that contains proteins and a lipid bilayera.

Life Span

One of the basic parameters of organism is its life span. Some animals live as short as one day, while some plants can live thousands of years. Aging is important when determining life span of most organisms, bacterium, a virus or even a prion.

Evolution

In biology, the theory of universal common descent proposes that all organisms on Earth are descended from a common ancestor or ancestral gene pool. Evidence for common descent may be found in traits shared between all living organisms. In Darwin's day, the evidence of shared traits was based solely on visible observation of morphologic similarities, such as the fact that all birds have wings, even those which do not fly. Today, there is strong evidence from genetics that all organisms have a common ancestor. For example, every living cell makes use of nucleic acids as its genetic material, and uses the same twenty amino acids as the building blocks for proteins. All organisms use the same genetic code (with some extremely rare and minor deviations) to translate nucleic acid sequences into proteins. The universality of these traits strongly suggests common ancestry, because the selection of many of these traits seems arbitrary.

Information about the early development of life includes input from the fields of geology and planetary science. These sciences provide information about the history of the Earth and the changes produced by life. However, a great deal of information about the early Earth has been destroyed by geological processes over the course of time.

Origin of Life

The chemical evolution from self-catalytic chemical reactions to life is not a part of biological evolution, but it is unclear at which point such increasingly complex sets of reactions became what we would consider, today, to be living organisms.

Not much is known about the earliest developments in life. However, all existing organisms share certain traits, including cellular structure and genetic code. Most scientists interpret this to mean all existing organisms share a common ancestor, which had already developed the most fundamental cellular processes, but there is no scientific consensus on the relationship of the three domains of life (Archaea, Bacteria, Eukaryota) or the origin of life. Attempts to shed light on the earliest history of life generally focus on the behavior of macromolecules, particularly RNA, and the behavior of complex systems. The emergence of oxygenic photosynthesis (around 3 billion years ago) and the subsequent emergence of an oxygen-rich, non-reducing atmosphere can be traced through the formation of banded iron deposits, and later red beds of iron oxides. This was a necessary prerequisite for the development of aerobic cellular respiration, believed to have emerged around 2 billion years ago.

In the last billion years, simple multicellular plants and animals began to appear in the oceans. Soon after the emergence of the first animals, the Cambrian explosion (a period of unrivaled and remarkable, but brief, organismal diversity documented in the fossils found at the Burgess Shale) saw the creation of all the major body plans, or phyla, of modern animals. This event is now believed to have been triggered by the development of the Hox genes. About 500 million years ago, plants and fungi colonized the land, and were soon followed by arthropods and other animals, leading to the development of land ecosystems with which we are familiar.

The evolutionary process may be exceedingly slow. Fossil evidence indicates that the diversity and complexity of modern life has developed over much of the history of the earth. Geological evidence indicates that the Earth is approximately 4.6 billion years old. Studies on guppies by David Reznick at the University of California, Riverside, however, have shown that the rate of evolution through natural selection can proceed 10 thousand to 10 million times faster than what is indicated in the fossil record.. Such comparative studies however are invariably biased by disparities in the time scales over which evolutionary change is measured in the laboratory, field experiments, and the fossil record.

Horizontal Gene Transfer, and the History of Life

The ancestry of living organisms has traditionally been reconstructed from morphology, but is increasingly supplemented with phylogenetics-the reconstructiion of phylogenies by the comparison of genetic (DNA) sequence.

"Sequence comparisons suggest recent horizontal transfer of many genes among diverse species including across the boundaries of phylogenetic 'domains'. Thus determining the phylogenetic history of a species can not be done conclusively by determining evolutionary trees for single genes." Biologist Gogarten suggests "the original metaphor of a tree no longer fits the data from recent genome research" therefore "biologists [should] use the metaphor of a mosaic to describe the different histories combined in individual genomes and use [the] metaphor of a net to visualize the rich exchange and cooperative effects of HGT among microbes."

ARCHEA

Archaea were identified in 1977 by Carl Woese and George E. Fox as being a separate branch based on their separation from other prokaryotes on 16S rRNA phylogenetic trees. These two groups were originally named the Archaebacteria and Eubacteria, treated as kingdoms or subkingdoms, which Woese and Fox termed Urkingdoms. Woese argued that they represented fundamentally different branches of living things. He later renamed the groups Archaea and Bacteria to emphasize this, and argued that together with Eukarya they compose three Domains of living organisms.

The biological term, Archaea, should not be confused with the geologic phrase Archean eon, also known as the Archeozoic era. This latter term refers to the primordial period of earth history when Archaea and Bacteria were the only cellular organisms living on the planet. Probable fossils of these microbes have been dated to almost 3.8 billion years ago (3800 mya).

Archaea, Bacteria and Eukaryotes

Archaea are similar to other prokaryotes in most aspects of cell structure and metabolism. However, their genetic transcription and translation-the two central processes in molecular biology-do not show many typical bacterial features, and are in many aspects similar to those of eukaryotes. For instance, archaean translation uses eukaryotic-like initiation and elongation factors, and their transcription involves TATA-binding proteins and TFIIB as in eukaryotes. Many archaeal tRNA and rRNA genes harbor unique archaeal introns which are neither like eukaryotic introns, nor like bacterial (type I and type II etc which can "home") introns. Several other characteristics also set the Archaea apart. Like bacteria and eukaryotes, archaea possess glycerol-based phospholipids. However, three features of the archaeal lipids are unusual:

- The archaeal lipids are unique because the stereochemistry of the glycerol is the reverse of that found in bacteria and eukaryotes. This is strong evidence for a different biosynthetic pathway.
- Most bacteria and eukaryotes have membranes composed mainly of glycerol-ester lipids, whereas archaea have membranes composed of glycerol-ether lipids. Even when bacteria have ether-linked lipids, the stereochemistry of the glycerol is the bacterial form. These differences may be an adaptation on the part of Archaea to hyperthermophily. However, it is worth noting that even mesophilic archaea have ether-linked lipids.
- Archaeal lipids are based upon the isoprenoid sidechain. This is a five-carbon unit that is also common in rubber and as a component of some vitamins common in bacteria and eukaryotes. However, only the archaea incorporate these compounds into their cellular lipids, frequently as C-20 (four monomers) or C-40 (eight monomers) side-chains. In some archaea, the C-40 isoprenoid side-chain is actually long enough to span the membrane, forming a monolayer for a cell membrane with glycerol phosphate moieties on both ends. Although dramatic, this adaptation is most common in the extremely thermophilic archaea.

Although not unique, the archaeal cell walls are also unusual. For instance, the cell walls of most archaea are formed by surface-layer proteins or an S-layer. S-layers are common in bacteria, where they serve as the sole cell-wall component in some organisms (like the Planctomyces) or an outer layer in many organisms with peptidoglycan. With the exception of one group of methanogens, archaea lack a peptidoglycan wall (and in the case of the exception, the peptidoglycan is very different from the type found in bacteria). Archaeans also have flagella that are notably different in composition and development from the superficially similar flagella of bacteria. Bacterial flagella is a modified type III secretion system, while archeal flagella resemble type IV pilli which use a sec dependent secretion system somewhat similar to but different from type II secretion system. Archea are single celled. They are prokaryotic, have no nucleus, and have one circular chromosome.

Habitats

Many archaeans are extremophiles. They can survive and thrive at even relatively high temperatures, often above 100°C, as found in geysers and black smokers. Others are found in very cold habitats or in highly-saline, acidic, or alkaline water. However, other archaeans are mesophiles, and have been found in environments like marshland, sewage, sea water and soil. Many methanogenic archaea are found in the digestive tracts of animals such as ruminants, termites, and humans. As of 2007, no clear examples of archaean pathogens are known, although a relationship has been proposed between the presence of some methanogens and human periodontal disease.

Archaea are usually placed into three groups based on preferred habitat. These are the halophiles, methanogens, and thermophiles. Halophiles live in extremely saline environments. Methanogens live in anaerobic environments and produce methane. These can be found in sediments or in the intestines of animals. Thermophiles live in places that have high temperatures, such as hot springs. These groups do not necessarily agree with molecular phylogenies, are not necessarily complete, nor are they mutually exclusive. Nonetheless, they are a useful starting point for more detailed studies.

Recently, several studies have shown that archaea exist not only in mesophilic and thermophilic environments but are also present, sometimes in high numbers, at low temperatures as well. It is increasingly becoming recognised that methanogens are commonly present in low-temperature environments such as cold sediments. Some studies have even suggested that at these temperatures the pathway by which methanogenesis occurs may change due to the thermodynamic constraints imposed by low temperatures. Perhaps even more significant are the large numbers of archaea found throughout most of the world's oceans, a predominantly cold environment (Giovannoni and Stingl, 2005). These archaea, which belong to several deeply branching lineages unrelated to those previously known, can be present in extremely high numbers (up to 40% of the microbial biomass) although almost none have been isolated in pure culture. Currently we have almost no information regarding the physiology of these organisms meaning that their effects on global biogeochemical cycles remain unknown. One recent study has shown, however, that one group of marine crenarchaeota are capable of nitrification, a trait previously unknown among the archaea.

Form

Individual archaeans range from 0.1 μm to over 15 μm in diameter, and some form aggregates or filaments up to 200 μm in length. They occur in various shapes, such as spherical, rod-shape, spiral, lobed, or rectangular. Recently, a species of flat, square archaean that lives in hypersaline pools has been discovered. They also exhibit a variety of different types of metabolism. Of note, the halobacteria can use light to produce ATP, although no Archaea conduct photosynthesis with an electron transport chain, as occurs in other groups. Rather light-activated ion pumps like bacteriorhodopsin and halorhodopsin play a role in generating ion gradients, whose energy then leads to production of ATP.

Evolution and Classification

Archaea are divided into two main groups based on rRNA trees, the Euryarchaeota and Crenarchaeota. Two other groups have been tentatively created for certain environmental samples and the peculiar species Nanoarchaeum equitans, discovered in 2002 by Karl Stetter, but their affinities are uncertain.

Woese argued that the bacteria, archaea, and eukaryotes each represent a primary line of descent that diverged early on from an ancestral progenote with poorly-developed genetic machinery. This hypothesis is reflected in the name Archaea, from the Greek archae or ancient. Later he treated these groups formally as domains, each comprising several kingdoms. This division has become very popular, although the idea of the progenote itself is not generally supported. Some biologists, however, have argued that the archaebacteria and eukaryotes arose from specialized eubacteria. The relationship between Archaea and Eukarya remains an important problem. Aside from the similarities noted above, many genetic trees group the two together. Some place eukaryotes closer to Eurarchaeota than Crenarchaeota are, although the membrane chemistry suggests otherwise.

However, the discovery of archaean-like genes in certain bacteria, such as Thermotoga, makes their relationship difficult to determine, as horizontal gene transfer may have occurred. Some have suggested that eukaryotes arose through fusion of an archaean and eubacterium, which became the nucleus and cytoplasm, which accounts for various genetic similarities but runs into difficulties explaining cell structure.

Single gene sequencing for systematics has led to whole genome sequencing; by January, 2007, 31 archaeal genomes have been completed with 29 partially completed .

BACTERIA

Bacteria (singular: bacterium) are unicellular microorganisms. They are typically a few micrometres long and have many shapes including spheres, rods, and spirals. The study of bacteria is bacteriology, a branch of microbiology. Bacteria are ubiquitous in every habitat on Earth, growing in soil, acidic hot springs, radioactive waste, seawater, and deep in the earth's crust. Some bacteria can even survive in the extreme cold and vacuum of outer space.

There are typically 40 million bacterial cells in a gram of soil and a million bacterial cells in a millilitre of fresh water; in all, there are approximately five nonillion (5×1030) bacteria in the world. Bacteria are vital in recycling nutrients, and many important steps in nutrient cycles depend on bacteria, such as the fixation of nitrogen from the atmosphere. However, most of these bacteria have not been characterised, and only about half of the phyla of bacteria have species that can be cultured in the laboratory.

There are approximately 10 times as many bacterial cells as human cells in the human body, with large numbers of bacteria on the skin and in the digestive tract. Although the vast majority of these bacteria are rendered harmless or beneficial by the protective effects of the immune system, a few pathogenic bacteria cause infectious diseases, including cholera, syphilis, anthrax, leprosy and bubonic plague.

The most common fatal bacterial diseases are respiratory infections, with tuberculosis alone killing about 2 million people a year, mostly in sub-Saharan Africa. In developed countries, antibiotics are used to treat bacterial infections and in various agricultural processes, so antibiotic resistance is becoming common. In industry, bacteria are important in processes such as wastewater treatment, the production of cheese and yoghurt, and the manufacture of antibiotics and other chemicals.

Bacteria are prokaryotes. Unlike animals and other eukaryotes, bacterial cells do not contain a nucleus or other membrane-bound organelles. Although the term bacteria traditionally included all prokaryotes, the scientific nomenclature changed after the discovery that prokaryotic life consists of two very different groups of organisms that evolved independently from an ancient common ancestor. These evolutionary domains are called Bacteria and Archaea.

MORPHOLOGY

Bacteria display a wide diversity of shapes and sizes, called morphologies. Bacterial cells are about 10 times smaller than eukaryotic cells and are typically 0.5-5.0 micrometres in length. However, a few species-for example Thiomargarita namibiensis and Epulopiscium fishelsoni-are up to half a millimetre long and are visible to the unaided eye. Among the smallest bacteria are members of the genus Mycoplasma, which measure only 0.3 micrometres, as small as the largest viruses.

Most bacterial species are either spherical, called cocci (sing. coccus, from Greek kókkos, grain, seed) or rod-shaped, called bacilli (sing. bacillus, from Latin baculus, stick). Some rod-shaped bacteria, called vibrio, are slightly curved or comma-shaped; others, can be spiral-shaped, called spirilla, or tightly coiled, called spirochetes. A small number of species even have tetrahedral or cuboidal shapes. This wide variety of shapes is determined by the bacterial cell wall and cytoskeleton, and is important because it can influence the ability of bacteria to acquire nutrients, attach to surfaces, swim through liquids and escape predators.

Many bacterial species exist simply as single cells; others associate in characteristic patterns: Neisseria form diploids (pairs), Streptococcus form chains, and Staphylococcus group together in "bunch of grapes" clusters. Bacteria can also be elongated to form filaments, for example the Actinobacteria. Filamentous bacteria are often surrounded by a sheath that contains many individual cells; certain species, such as the genus Nocardia, form complex, branched filaments, similar in appearance to fungal mycelia.

Bacteria often attach to surfaces and form dense aggregations called biofilms or microbial mats. These films can range from a few micrometers in thickness to up to half a metre in depth, and may contain multiple species of bacteria, protists and archaea. Bacteria living in biofilms display a complex arrangement of cells and extracellular components, forming secondary structures such as microcolonies, through which there are networks of channels to enable better diffusion of nutrients. In natural environments, such as soil or the surfaces of plants, the majority of bacteria are bound to surfaces in biofilms. Biofilms are also important for chronic bacterial infections and infections of implanted medical devices, as bacteria protected within these structures are much harder to kill than individual bacteria.

Even more complex morphological changes are sometimes possible. For example, when starved of amino acids, Myxobacteria detect surrounding cells in a process known as quorum sensing, migrate towards each other, and aggregate to form fruiting bodies up to 500 micrometres long and containing approximately 100,000 bacterial cells. In these fruiting bodies, the bacteria perform separate tasks; this type of cooperation is a simple type of multicellular organisation. For example, about one in 10 cells migrate to the top of these fruiting bodies and differentiate into a specialised dormant state called myxospores, which are more resistant to desiccation and other adverse environmental conditions than are ordinary cells.

CELLULAR STRUCTURE

Intracellular Structures

The bacterial cell is surrounded by a lipid membrane, or cell membrane, which encompasses the contents of the cell and acts as a barrier to hold nutrients, proteins and other essential components of the cytoplasm within the cell. As they are prokaryotes, bacteria do not have membrane-bound organelles in their cytoplasm and thus contain few intracellular structures. They consequently lack a nucleus, mitochondria, chloroplasts and the other organelles present in eukaryotic cells, such as the Golgi apparatus and endoplasmic reticulum.

Many important biochemical reactions, such as energy generation, occur due to concentration gradients across membranes, creating a potential difference analogous to a battery. The absence of internal membranes in bacteria means these reactions, such as electron transport, occur across the cell membrane, between the cytoplasm and the periplasmic space.

Bacteria do not have a membrane-bound nucleus, and their genetic material is typically a single circular chromosome located in the cytoplasm in an irregularly shaped body called the nucleoid. The nucleoid contains the chromosome with associated proteins and RNA. Like all living organisms, bacteria contain ribosomes for the production of proteins, but the structure of the bacterial ribosome is different from those of eukaryotes and Archaea.

The order Planctomycetes are an exception to the general absence of internal membranes in bacteria, because they have a membrane around their nucleoid and contain other membrane-bound cellular structures.

Some bacteria produce intracellular nutrient storage granules, such as glycogen, polyphosphate, sulfur or polyhydroxyalkanoates. These granules enable bacteria to store compounds for later use. Certain bacterial species, such as the photosynthetic Cyanobacteria, produce internal gas vesicles, which they use to regulate their buoyancy to achieve optimal light intensity and/ or nutrient levels.

Extracellular Structures

Around the outside of the cell membrane is the bacterial cell wall. Bacterial cell walls are made of peptidoglycan (called murein in older sources), which is made from polysaccharide chains cross-linked by unusual peptides containing D-amino acids. Bacterial cell walls are different from the cell walls of plants and fungi, which are made of cellulose and chitin, respectively. The cell wall of bacteria is also distinct from that of Archaea, which do not contain peptidoglycan. The cell wall is essential to the survival of many bacteria, and the antibiotic penicillin is able to kill bacteria by inhibiting a step in the synthesis of peptidoglycan.

There are broadly speaking two different types of cell wall in bacteria, called Gram-positive and Gram-negative. The names originate from the reaction of cells to the Gram stain, a test long-employed for the classification of bacterial species.

Gram-positive bacteria possess a thick cell wall containing many layers of peptidoglycan and teichoic acids. In contrast, Gram-negative bacteria have a relatively thin cell wall consisting of a few layers of peptidoglycan surrounded by a second lipid membrane containing lipopolysaccharides and lipoproteins. Most bacteria have the Gram-negative cell wall, and only the Firmicutes and Actinobacteria (previously known as the low G+C and high G+C Gram-positive bacteria, respectively) have the alternative Gram-positive arrangement.

These differences in structure can produce differences in antibiotic susceptibility; for instance, vancomycin can kill only Gram-positive bacteria and is ineffective against Gram-negative pathogens, such as Haemophilus influenzae or Pseudomonas aeruginosa.

In many bacteria an S-layer of rigidly-arrayed protein molecules covers the outside of the cell. This layer provides chemical and physical protection for the cell surface and can act as a macromolecular diffusion barrier. S-layers have diverse but mostly poorly-understood functions, but are known to act as virulence factors in Campylobacter and contain surface enzymes in Bacillus stearothermophilus.

Flagella are rigid protein structures, about 20 nanometres in diameter and up to 20 micrometres in length, that are used for motility. Flagella are driven by the energy released by the transfer of ions down an electrochemical gradient across the cell membrane.

Fimbriae are fine filaments of protein, just 2-10 nanometres in diameter and up to several micrometers in length. They are distributed over the surface of the cell, and resemble fine hairs when seen under the electron microscope. Fimbriae are believed to be involved in attachment to solid surfaces or to other cells and are essential for the virulence of some bacterial pathogens. Pili (sing. pilus) are cellular appendages, slightly larger than fimbriae, that can transfer genetic material between bacterial cells in a process called conjugation.

Capsules or slime layers are produced by many bacteria to surround their cells, and vary in structural complexity: ranging from a disorganised slime layer of extra-cellular polymer, to a highly structured capsule or glycocalyx. These structures can protect cells from engulfment by eukaryotic cells, such as macrophages. They can also act as antigens and be involved in cell recognition, as well as aiding attachment to surfaces and the formation of biofilms.

The assembly of these extracellular structures is dependent on bacterial secretion systems. These transfer proteins from the cytoplasm into the periplasm or into the environment around the cell. Many types of secretion systems are known and these structures are often essential for the virulence of pathogens, so are intensively studied.

Endospores

Certain genera of Gram-positive bacteria, such as Bacillus, Clostridium, Sporohalobacter, Anaerobacter and Heliobacterium, can form highly-resistant, dormant structures called endospores. In almost all cases, one endospore is formed and this is not a reproductive process, although Anaerobacter can make up to seven endospores in a single cell. Endospores have a central core of cytoplasm containing DNA and ribosomes surrounded by a cortex layer and protected by an impermeable and rigid coat.

Endospores show no detectable metabolism and can survive extreme physical and chemical stresses, such as high levels of UV light, gamma radiation, detergents, disinfectants, heat, pressure and desiccation. In this dormant state, these organisms may remain viable for millions of years, and endospores even allow bacteria to survive exposure to the vacuum and radiation in space. Endospore-forming bacteria can also cause disease: for example, anthrax can be contracted by the inhalation of Bacillus anthracis endospores, and contamination of deep puncture wounds with Clostridium tetani endospores causes tetanus.

Metabolism

In contrast to higher organisms, bacteria exhibit an extremely wide variety of metabolic types. The distribution of metabolic traits within a group of bacteria has traditionally been used to define their taxonomy, but these traits often do not correspond with modern genetic classifications. Bacterial metabolism is classified on the basis of three major criteria: the kind of energy used for growth, the source of carbon, and the electron donors used for growth. An additional criterion of respiratory microorganisms are the electron acceptors used for aerobic or anaerobic respiration.

Carbon metabolism in bacteria is either heterotrophic, where organic carbon compounds are used as carbon sources, or autotrophic, meaning that cellular carbon is obtained by fixing carbon dioxide. Typical autotrophic bacteria are phototrophic cyanobacteria, green sulfur-bacteria and some purple bacteria, but also many chemolithotrophic species, e. g. nitrifying or sulfur-oxidising bacteria. Energy metabolism of bacteria is either based on phototrophy, the use of light through photosynthesis, or on chemotrophy, the use of chemical substances for energy, which are mostly oxidised at the expense of oxygen or alternative electron acceptors (aerobic/anaerobic respiration).

Finally, bacteria are further divided into lithotrophs that use inorganic electron donors and organotrophs that use organic compounds as electron donors. Chemotrophic organisms use the respective electron donors for energy conservation (by aerobic/

anaerobic respiration or fermentation) and biosynthetic reactions (e.g. carbon dioxide fixation), whereas phototrophic organisms use them only for biosynthetic purposes. Respiratory organisms use chemical compounds as a source of energy by taking electrons from the reduced substrate and transferring them to a terminal electron acceptor in a redox reaction. This reaction releases energy that can be used to synthesise ATP and drive metabolism. In aerobic organisms, oxygen is used as the electron acceptor. In anaerobic organisms other inorganic compounds, such as nitrate, sulfate or carbon dioxide are used as electron acceptors. This leads to the ecologically-important processes of denitrification, sulfate reduction and acetogenesis, respectively.

Another way of life of chemotrophs in the absence of possible electron acceptors is fermentation, where the electrons taken from the reduced substrates are transferred to oxidised intermediates to generate reduced fermentation products (e.g. lactate, ethanol, hydrogen, butyrate). Fermentation is possible, because the energy content of the substrates is higher than that of the products, which allows the organisms to synthesise ATP and drive their metabolism.

These processes are also important in biological responses to pollution; for example, sulfate-reducing bacteria are largely responsible for the production of the highly toxic forms of mercury (methyl-and dimethylmercury) in the environment. Non-respiratory anaerobes use fermentation to generate energy and reducing power, secreting metabolic by-products (such as ethanol in brewing) as waste. Facultative anaerobes can switch between fermentation and different terminal electron acceptors depending on the environmental conditions in which they find themselves.

Lithotrophic bacteria can use inorganic compounds as a source of energy. Common inorganic electron donors are hydrogen, carbon monoxide, ammonia (leading to nitrification), ferrous iron and other reduced metal ions, and several reduced sulfur compounds. Unusually, the gas methane can be used by methanotrophic bacteria as both a source of electrons and a substrate for carbon anabolism. In both aerobic phototrophy and chemolithotrophy, oxygen is used as a terminal electron acceptor, while under anaerobic conditions inorganic compounds are used instead. Most lithotrophic organisms are autotrophic, whereas organotrophic organisms are heterotrophic.

In addition to fixing carbon dioxide in photosynthesis, some bacteria also fix nitrogen gas (nitrogen fixation) using the enzyme nitrogenase. This environmentally important trait can be found in bacteria of nearly all the metabolic types listed above, but is not universal.

Growth and Reproduction

Unlike multicellular organisms, increases in the size of bacteria (cell growth) and their reproduction by cell division are tightly linked in unicellular organisms. Bacteria grow to a fixed size and then reproduce through binary fission, a form of asexual reproduction. Under optimal conditions, bacteria can grow and divide extremely rapidly, and bacterial populations can double as quickly as every 9.8 minutes. In cell division, two identical clone daughter cells are produced. Some bacteria, while still reproducing asexually, form more complex reproductive structures that facilitate the dispersal of the newly-formed daughter cells. Examples include fruiting body formation by Myxobacteria and arial hyphae formation by Streptomyces, or budding. Budding involves a cell forming a protrusion that breaks away and produces a daughter cell. In the laboratory, bacteria are usually grown using solid or liquid media. Solid growth media such as agar plates are used to isolate pure cultures of a bacterial strain. However, liquid growth media are used when measurement of growth or large volumes of cells are required. Growth in stirred liquid media occurs as an even cell suspension, making the cultures easy to divide and transfer, although isolating single bacteria from liquid media is difficult. The use of selective media (media with specific nutrients added or deficient, or with antibiotics added) can help identify specific organisms.

Most laboratory techniques for growing bacteria use high levels of nutrients to produce large amounts of cells cheaply and quickly. However, in natural environments nutrients are limited, meaning that bacteria cannot continue to reproduce indefinitely. This nutrient limitation has led the evolution of different growth strategies. Some organisms can grow extremely rapidly when nutrients become available, such as the formation of algal (and cyanobacterial) blooms that often occur in lakes during the summer. Other organisms have adaptations to harsh environments, such as the production of multiple antibiotics by Streptomyces that inhibit the growth of competing microorganisms. In nature, many organisms live in communities (e.g. biofilms) which may allow for increased supply of nutrients and protection from environmental stresses. These relationships can be essential for growth of a particular organism or group of organisms (syntrophy).

Bacterial growth follows three phases. When a population of bacteria first enter a high-nutrient environment that allows growth, the cells need to adapt to their new environment. The first phase of growth is the lag phase, a period of slow growth when the cells are adapting to fast growth. The lag phase has high biosynthesis rates, as enzymes and nutrient transporters are produced. The second phase of growth is the logarithmic phase (log phase), also known as the exponential phase. The log phase is marked by rapid exponential growth. The rate at which cells grow during this phase is known as the growth rate (k), and the time it takes the cells to double is known as the generation time (g). During log phase, nutrients are metabolised at maximum speed until one of the nutrients is depleted and starts limiting growth. The final phase of growth is the stationary phase and is caused by depleted nutrients. The cells reduce their metabolic activity and consume non-essential cellular proteins. The stationary phase is a transition from rapid growth to a stress response state and there is increased expression of genes involved in DNA repair, antioxidant metabolism and nutrient transport.

Bacterial Genetics

Most bacteria have a single circular chromosome that can range in size from only 580,000 base pairs in the human pathogen Mycoplasma genitalium, to 12,200,000 base pairs in the soil-dwelling bacteria Sorangium cellulosum. Spirochaetes are a notable exception to this arrangement, with bacteria such as Borrelia burgdorferi, the cause of Lyme disease, containing a single linear chromosome. Bacteria may also contain plasmids, which are small extra-chromosomal DNAs that may contain genes for antibiotic resistance or virulence factors. Another type of bacterial DNA are integrated viruses (bacteriophages). Many types of bacteriophage exist, some simply infect and lyse their host bacteria, while others insert into the bacterial chromosome. A bacteriophage can contain genes that contribute to its host's phenotype: for example, in the evolution of Escherichia coli O157:H7 and Clostridium botulinum, the toxin genes in an integrated phage converted a harmless ancestral bacteria into a lethal pathogen.

Bacteria, as asexual organisms, inherit identical copies of their parent's genes (i.e., they are clonal). However, all bacteria can evolve by selection on changes to their genetic material DNA caused by genetic recombination or mutations. Mutations come from errors made during the replication of DNA or from exposure to mutagens. Mutation rates vary widely among different species of bacteria and even among different clones of a single species of bacteria. Genetic changes in bacterial genomes come from either random mutation during replication or "stress-directed mutation", where genes involved in a particular growth-limiting process have an increased mutation rate.

Some bacteria also transfer genetic material between cells. This can occur in three main ways. Firstly, bacteria can take up exogenous DNA from their environment, in a process called transformation. Often, the genes transferred are not from within the main bacterial chromosome, but are carried on a small circular piece of DNA called a plasmid. Genes can also be transferred by the process of transduction, when the integration of a bacteriophage introduces foreign DNA into the chromosome. The third method of gene transfer is bacterial conjugation, where DNA is transferred through direct cell contact. This gene acquisition from other bacteria or the environment is called horizontal gene transfer and may be common under natural conditions. Gene transfer is particularly important in antibiotic resistance as it allows the rapid transfer of resistance genes between different pathogens.

Movement

Motile bacteria can move using flagella, bacterial gliding, twitching motility or changes of buoyancy. A unique group of bacteria, the spirochaetes, have structures similar to flagella, called axial filaments that are found between two membranes in the periplasmic space. They have a distinctive helical body that twists about as it moves. In twitching motility, bacterial use their type IV pili as a grappling hook, repeatedly extending it, anchoring it and then retracting it with remarkable force.

Bacterial species differ in the number and arrangement of flagella on their surface; some have a single flagellum (monotrichous), a flagellum at each end (amphitrichous), clusters of flagella at the poles of the cell (lophotrichous), while others have flagella distributed over the entire surface of the cell (peritrichous). The bacterial flagella is the best-understood motility structure in any organism and is made of about 20 proteins, with approximately another 30 proteins required for its regulation and assembly. The flagellum is a rotating structure driven by a motor at the base that uses the proton-motive force for power. This motor drives the motion of the filament, which acts as a propeller. Many bacteria (such as E. coli) have two distinct modes of movement: forward movement (swimming) and tumbling. The tumbling allows them to reorient and makes their movement a three-dimensional random walk.

Motile bacteria are attracted or repelled by certain stimuli in behaviors called taxes: these include chemotaxis, phototaxis and magnetotaxis. In one peculiar group, the myxobacteria, individual bacteria move together to form waves of cells that then differentiate to form fruiting bodies containing spores. The myxobacteria move only when on solid surfaces, unlike E. coli which is motile in liquid or solid media.

Several Listeria and Shigella species move inside host cells by usurping the cytoskeleton, which is normally used to move organelles inside the cell. By promoting actin polymerization at one pole of their cells, they can form a kind of tail that pushes them through the host cell's cytoplasm.

Classification and Identification

Classification seeks to describe the diversity of bacterial species by naming and grouping organisms based on similarities. Bacteria can be classified on the basis of cell structure, cellular metabolism or on differences in cell components such as DNA, fatty acids, pigments, antigens and quinones. While these schemes allowed the identification and classification of bacterial strains, it was unclear whether these differences represented variation between distinct species or between strains of the same species.

This uncertainty was due to the lack of distinctive structures in most bacteria, as well as lateral gene transfer between unrelated species. Due to lateral gene transfer, some closely-related bacteria can have very different morphologies and metabolisms. To overcome this uncertainty, modern bacterial classification emphasizes molecular systematics, using genetic techniques such as guanine cytosine ratio determination, genome-genome hybridization, as well as sequencing genes that have not undergone extensive lateral gene transfer, such as the rRNA gene.

The term "bacteria" was traditionally applied to all microscopic, single-celled prokaryotes. However, molecular systematics showed prokaryotic life to consist of two separate domains, originally called Eubacteria and Archaebacteria, but now called Bacteria and Archaea that evolved independently from an ancient common ancestor. These two domains, along with Eukarya, are the basis of the three-domain system, which is currently the most widely-used classification system in bacteriology. However, due to the relatively recent introduction of molecular systematics and the analysis of genome sequences, bacterial classification remains a changing and expanding field. For example, a few biologists argue that Archaea evolved from Gram-positive bacteria.

Identification of bacteria in the laboratory is particularly relevant in medicine, where the correct treatment is determined by the bacterial species causing an infection. Consequently, the need to identify human pathogens was a major impetus for the development of techniques to identify bacteria. The Gram stain, developed in 1884 by Hans Christian Gram, characterises bacteria based on the structural characteristics of their cell walls. The thick layers of peptidoglycan in the "Gram-positive" cell wall stain purple, while the thin "Gram-negative" cell wall appears pink. By combining morphology and Gram-staining, most bacteria can be classified as belonging to one of four groups (Gram-positive cocci, Gram-positive bacilli, Gram-negative cocci and Gram-negative bacilli). Some organisms are best identified by stains other than the Gram stain, particularly mycobacteria or Nocardia, which show acid-fastness on Ziehl-Neelsen or similar stains. Other organisms may need to be identified by their growth in special media, or by other techniques, such as serology.

Culture techniques are designed to promote the growth and identify particular bacteria, while restricting the growth of the other bacteria in the sample. Often these techniques are designed for specific specimens; for example, a sputum sample will be treated to identify organisms that cause pneumonia, while stool specimens are cultured on selective media to identify organisms that cause diarrhoea, while preventing growth of non-pathogenic bacteria. Specimens that are normally sterile, such as blood, urine or spinal fluid, are cultured under conditions designed to grow all possible organisms. Once a pathogenic organism has been isolated, it can be further characterised by its morphology, growth patterns such as (aerobic or anaerobic growth, patterns of hemolysis) and staining.

As with bacterial classification, identification of bacteria is increasingly using molecular methods. Diagnostics using such DNA-based tools, such as polymerase chain reaction, are increasingly popular due to their specificity and speed, compared to culture-based methods. These methods also allow the detection and identification of "viable but nonculturable" cells that are metabolically-active but non-dividing.

Interactions with Other Organisms

Despite their apparent simplicity, bacteria can form complex associations with other organisms. These symbiotic associations can be divided into parasitism, mutualism and commensalism. Due to their small size, commensal bacteria are ubiquitous and grow on animals and plants exactly as they will grow on any other surface. However, their growth can be increased by warmth and sweat, and large populations of these organisms in humans are the cause of body odor.

Mutualists

Certain bacteria form close spatial associations that are essential for their survival. One such mutualistic association, called interspecies hydrogen transfer, occurs between clusters of anaerobic bacteria that consume organic acids such as butyric acid or propionic acid and produce hydrogen, and methanogenic Archaea that consume hydrogen. The bacteria in this association are unable to consume the organic acids as this reaction produces hydrogen that accumulates in their surroundings. Only the intimate association with the hydrogen-consuming Archaea keeps the hydrogen concentration low enough to allow the bacteria to grow.

In soil, microorganisms which reside in the rhizosphere (a zone that includes the root surface and the soil that adheres to the root after gentle shaking) carry out nitrogen fixation, converting nitrogen gas to nitrogenous compounds. This serves to provide an easily absorbable form of nitrogen for many plants, which cannot fix nitrogen themselves. Many other bacteria are found as symbionts in humans and other organisms. For example, the presence of over 1,000 bacterial species in the normal human gut flora of the intestines can contribute to gut immunity, synthesise vitamins such as folic acid, vitamin K and biotin, as well as fermenting complex undigestible carbohydrates. Bacteria that offer some benefit to human hosts include Lactobacillus species, which convert milk protein to lactic acid in the gut. The presence of this gut flora also inhibits the growth of potentially pathogenic bacteria (usually through competitive exclusion) and these beneficial bacteria are consequently sold as probiotic dietary supplements.

Pathogens

If bacteria form a parasitic association with other organisms, they are classed as pathogens. Pathogenic bacteria are a major cause of human death and disease and cause infections such as tetanus, typhoid fever, diphtheria, syphilis, cholera, food-borne illness, leprosy and tuberculosis. A pathogenic cause for a known medical disease may only be discovered many years after, as was the case with Helicobacter pylori and peptic ulcer disease. Bacterial diseases are also important in agriculture, with bacteria causing leaf spot, fireblight and wilts in plants, as well as Johne's disease, mastitis, salmonella and anthrax in farm animals.

Each species of pathogen has a characteristic spectrum of interactions with its human hosts. Some organisms, such as Staphylococcus or Streptococcus, can cause skin infections, pneumonia, meningitis and even overwhelming sepsis, a systemic inflammatory response producing shock, massive vasodilation and death. Yet these organisms are also part of the normal human flora and usually exist on the skin or in the nose without causing any disease at all. Other organisms invariably cause disease in humans, such as the Rickettsia, which are obligate intracellular parasites able to grow and reproduce only within the cells of other organisms. One species of Rickettsia causes typhus, while another causes Rocky Mountain spotted fever. Chlamydia, another phylum of obligate intracellular parasites, contains species that can cause pneumonia, or urinary tract infection and may be involved in coronary heart disease. Finally, some species such as Pseudomonas aeruginosa, Burkholderia cenocepacia, and Mycobacterium avium are opportunistic pathogens and cause disease mainly in people suffering from immunosuppression or cystic fibrosis.

Bacterial infections may be treated with antibiotics, which are classified as bacteriocidal if they kill bacteria, or bacteriostatic if they just prevent bacterial growth. There are many types of antibiotics and each class inhibits a process that is different in the pathogen from that found in the host. An example of how antibiotics produce selective toxicity are chloramphenicol and puromycin, which inhibit the bacterial ribosome, but not the structurally-different eukaryotic ribosome. Antibiotics are used both in treating human disease and in intensive farming to promote animal growth, where they may be contributing to the rapid development of antibiotic resistance in bacterial populations. Infections can be prevented by antiseptic measures such as sterilizating the skin prior to piercing it with the needle of a syringe, and by proper care of indwelling catheters. Surgical and dental instruments are also sterilized to prevent contamination and infection by bacteria. Disinfectants such as bleach are used to kill bacteria or other pathogens on surfaces to prevent contamination and further reduce the risk of infection.

Significance in Technology and Industry

Bacteria, often Lactobacillus in combination with yeasts and molds, have been used for thousands of years in the preparation of fermented foods such as cheese, pickles, soy sauce, sauerkraut, vinegar, wine and yoghurt.

The ability of bacteria to degrade a variety of organic compounds is remarkable and has been used in waste processing and bioremediation. Bacteria capable of digesting the hydrocarbons in petroleum are often used to clean up oil spills. Fertilizer was added to some of the beaches in Prince William Sound in an attempt to promote the growth of these naturally occurring bacteria after the infamous 1989 Exxon Valdez oil spill. These efforts were effective on beaches that were not too thickly covered in oil. Bacteria are also used for the bioremediation of industrial toxic wastes. In the chemical industry, bacteria are most important in the production of enantiomerically pure chemicals for use as pharmaceuticals or agrichemicals.

Bacteria can also be used in the place of pesticides in the biological pest control. This commonly involves Bacillus thuringiensis (also called BT), a Gram-positive, soil dwelling bacterium. Subspecies of this bacteria are used as a Lepidopteran-specific insecticides under trade names such as Dipel and Thuricide. Because of their specificity, these pesticides are regarded as environmentally friendly, with little or no effect on humans, wildlife, pollinators and most other beneficial insects.

Because of their ability to quickly grow and the relative ease with which they can be manipulated, bacteria are the workhorses for the fields of molecular biology, genetics and biochemistry. By making mutations in bacterial DNA and examining the resulting phenotypes, scientists can determine the function of genes, enzymes and metabolic pathways in bacteria, then apply this knowledge to more complex organisms. This aim of understanding the biochemistry of a cell reaches its most complex expression in the synthesis of huge amounts of enzyme kinetic and gene expression data into mathematical models of entire organisms. This is achievable in some well-studied bacteria, with models of Escherichia coli metabolism now being produced and tested. This understanding of bacterial metabolism and genetics allows the use of biotechnology to bioengineer bacteria for the production of therapeutic proteins, such as insulin, growth factors, or antibodies.

BACTERIA IN THE HUMAN BODY

The human body contains a large number of bacteria, most of them performing tasks that are useful or even essential to human survival. Those that are expected to be present, and that under normal circumstances do not cause disease, are termed normal flora.

It is estimated that 500 to 1000 different species of bacteria live in the human body (Sears, 2005). Bacterial cells are much smaller than human cells, and there are about ten times as many bacteria as human cells in the body (1000 trillion (1015) versus 100 trillion (1014); Sears, 2005). Though normal flora are found on all surfaces exposed to the environment (on the skin and eyes, in the mouth, nose, small intestine, and colon), the vast majority of bacteria live in the large intestine.

Many of the bacteria in the digestive tract, collectively referred to as gut flora, are able to break down certain nutrients such as carbohydrates that humans otherwise could not digest. The majority of these commensal bacteria are anaerobes, meaning they survive in an environment with no oxygen.

Many of the bacteria of the normal flora can act as opportunistic pathogens at times of lowered immunity.

Escherichia coli is a bacterium that lives in the colon; it is an extensively studied model organism and probably the best understood organism of all. A number of types of bacteria, such as Actinomyces viscosus and A. naeslundii, live in the mouth, and constitute a sticky substance called plaque. If this is not removed by brushing, it hardens into calculus (also called tartar). The same bacteria also secrete acids that dissolve the tooth enamel, causing tooth decay.

The vaginal microflora consist mostly of various lactobacillus species. It was long thought that the most common of these species was Lactobacillus acidophilus, but it has later been shown that the most common one is L. iners followed by L. crispatus. Other lactobacilli found in the vagina are L. delbruekii and L. gasseri. Disturbance of the vaginal flora can lead to bacterial vaginosis.

Bacteria and Human Health

Bacteria are vital for the maintenance of human health, but they also pose and a significant health threat by causing disease. There are 10 times more bacterial cells than human cells in the human body, with large numbers of bacteria on the skin and in the digestive tract. Due to their small size, commensal bacteria are ubiquitous and grow on animals and plants exactly as they will grow on any other surface. However, their growth can be increased by warmth and sweat and large populations of these organisms in humans are the cause of body odor. The presence of over 1,000 bacterial species in the normal human gut flora of the intestines can contribute to gut immunity, synthesise vitamins such as folic acid, vitamin K and biotin, as well as fermenting complex undigestable carbohydrates. Bacteria that offer some benefit to human hosts include Lactobacillus species, which convert milk protein to lactic acid in the gut. The presence of such bacterial colonies also inhibits the growth of potentially pathogenic bacteria (usually through competitive exclusion) and these beneficial bacteria are consequently sold as probiotic dietary supplements.

Although the vast majority of bacteria are harmless or beneficial, a few pathogenic bacteria cause infectious diseases. The most common bacterial disease is tuberculosis, caused by the bacterium Mycobacterium tuberculosis, which kills about 2 million people a year, mostly in sub-Saharan Africa. Pathogenic bacteria contribute to other globally important diseases, such as pneumonia, which can be caused by bacteria such as Streptococcus and Pseudomonas, and foodborne illnesses, which can be caused by bacteria such as Shigella, Campylobacter and Salmonella. Pathogenic bacteria also cause infections such as tetanus, typhoid fever, diphtheria, syphilis and leprosy.

Koch's postulates, proposed by Robert Koch in 1890, are criteria designed to establish a causal relationship between a causative microbe and a disease. A pathogenic cause for a known medical disease may only be discovered many years after, as was the case with Helicobacter pylori and peptic ulcer disease.

Each species of pathogen has a characteristic spectrum of interactions with its human hosts. Some organisms, such as Staphylococcus or Streptococcus, can cause skin infections, pneumonia, meningitis and even overwhelming sepsis, a systemic inflammatory response producing shock, massive vasodilation and death. Yet these organisms are also part of the normal human flora and usually exist on the skin or in the nose without causing any disease at all. Other organisms invariably cause disease in humans, such as the Rickettsia, which are obligate intracellular parasites able to grow and reproduce only within the cells of other organisms. One species of Rickettsia causes typhus, while another causes Rocky Mountain spotted fever.

Chlamydia, another phylum of obligate intracellular parasites, contains species that can cause pneumonia, or urinary tract infection and may be involved in coronary heart disease. Finally, some species, such as Pseudomonas aeruginosa, Burkholderia cenocepacia, and Mycobacterium avium, are opportunistic pathogens and cause disease mainly in people suffering from immunosuppression or cystic fibrosis.

Bacterial infections may be treated with antibiotics, which are classified as bacteriocidal if they kill bacteria, or bacteriostatic if they just prevent bacterial growth. There are many types of antibiotics and each class inhibits a process that is different in the pathogen from that found in the host. An example of how antibiotics produce selective toxicity are chloramphenicol and puromycin, which inhibit the bacterial ribosome, but not the structurally-different eukaryotic ribosome. Antibiotics are used both in treating human disease and in intensive farming to promote animal growth, where they may be contributing to the rapid development of antibiotic resistance in bacterial populations.

Infections can be prevented by antiseptic measures such as sterilizating the skin prior to piercing it with the needle of a syringe, and by proper care of indwelling catheters. Surgical and dental instruments are also sterilized to prevent contamination and infection by bacteria. Disinfectants such as bleach are used to kill bacteria or other pathogens on surfaces to prevent contamination and further reduce the risk of infection.

ENDOSYMBIONT

An endosymbiont is any organism that lives within the body or cells of another organism, i.e. forming an endosymbiosis (Greek: endo = inner, together and biosis = living). Examples are nitrogen-fixing bacteria (called rhizobia) which live in root nodules on legume roots, single-celled algae inside reef-building corals, and bacterial endosymbionts provide essential nutrients to about 10%-15% of insects.

Many instances of endosymbiosis are obligate, that is neither the endosymbiont nor the host can survive without the other, such as gutless marine worms, which get nutrition from their endosymbiotic bacteria. However, not all endosymbioses are obligate. Also, some endosymbioses can be harmful to either of the organisms involved.

It is generally agreed that certain organelles of the eukaryotic cell, especially mitochondria and plastids such as chloroplasts, originated as bacterial endosymbionts. This theory is called the endosymbiotic theory, confirmed and popularized by Lynn Margulis.

The Endosymbiont Theory and Mitochondria and Chloroplasts

The endosymbiont theory attempts to explain the origins of organelles such as mitochondria and chloroplasts in eukaryotic cells. The theory proposes that chloroplasts and mitochondria evolved from certain types of bacteria that prokaryotic cells engulfed through endophagocytosis. These cells and the bacteria trapped inside them entered a symbiotic relationship, a close association between different types of organisms over an extended time. However, more specifically, the relationship was endosymbiotic, meaning that one of the organisms (the bacteria) lived within the other (the prokaryotic cells).

According to the endosymbiont theory, an anaerobic cell probably ingested an aerobic bacterium but failed to digest it. The aerobic bacterium flourished within the cell because the cell's cytoplasm was abundant in half-digested food molecules. The bacterium digested these molecules with oxygen and gained great amounts of energy. Because the bacterium had so much energy, it probably leaked some of it as ATP into the cell's cytoplasm. This benefited the anaerobic cell because it enabled it to digest food aerobically. Eventually, the aerobic bacterium could no longer live independently from the cell, and it therefore became a mitochondrion.

The origin of the chloroplast is very similar to that of the mitochondrion. A cell must have captured a photosynthetic cyanobacterium and failed to digest it. The cyanobacterium thrived in the cell and eventually evolved into the first chloroplast. Other eukaryotic organelles may have also evolved through endosymbiosis. Scientists believe that cilia, flagella, centrioles, and microtubules may have come from a symbiosis between a spirilla-like bacterium and an early eukaryotic cell.

There are several examples of evidence that support the endosymbiont theory. Mitochondria and chloroplasts contain their own small supply of DNA, which may be remnants of the genome the organelles had when they were independent aerobic bacteria. The single most convincing evidence of the descent of organelles from bacteria is the position of mitochondria and plastid DNA sequences in phylogenetic trees of bacteria. Mitochondria have sequences that clearly indicate origin from a group of bacteria called the alpha-Proteobacteria. Plastids have DNA sequences that indicate origin from the cyanobacteria (blue-green algae). In addition, there are organisms alive today, called living intermediates, that are in a similar endosymbiotic condition to the prokaryotic cells and the aerobic bacteria. Living intermediates show that the evolution proposed by the endosymbiont theory is possible. For example, the giant amoeba Pelomyxa lacks mitochondria but has aerobic bacteria that carry out a similar role. A variety of corals, clams, snails, and one species of Paramecium permanently host algae in their cells. Many of the insect endosymbionts have been shown to have ancient associations with their hosts, involving strictly vertical inheritance. In addition, these insect symbionts have similar patterns of genome evolution to those found in true organelles: genome reduction, rapid rates of gene evolution, and bias in nucleotide base composition favoring adenine and thymine, at the expense of guanine and cytosine.

Further evidence of endosymbiosis are the prokaryotic ribosomes found within chloroplasts and mitochondria as well as the double membrane enclosing them. The inner membrane is thought to be the original membrane of the once independent prokaryote, while the outer one is thought to be the food vacuole it was enclosed in initially. Triple or quadruple membranes are found among certain algae, probably resulting from repeated endosymbiosis (although little else was retained of the engulfed cell). These modern organisms with endosymbiotic relationships with aerobic bacteria have verified the endosymbiotic theory, which explains the origin of mitochondria and chloroplasts from bacteria. Researchers in molecular and evolutionary biology no longer question this theory, although some of the details, such as the mechanisms for loss of genes from organelles to host nuclear genomes, are still being worked out.

Bacterial Endosymbionts in Marine Oligochaetes

Some marine oligochaeta (e.g Olavius or Inanidrillus) have obligate extracellular endosymbionts that fill the entire body of their host. These marine worms are nutritionally dependent on their symbiotic chemoautotrophic bacteria lacking any digestive or excretory system (no gut, mouth or nephridia).

Bacterial Endosymbionts in Other Marine Invertebrates

Extracellular endosymbionts are also represented in all 5 extant classes of Echinodermata (Crinoidea, Ophiuroidea, Asteroidea, Echinoidea, and Holothuroidea). Little is known of the nature of the association (mode of infection, transmission, metabolic requirements, etc.) but phylogenetic analysis indicates that these symbionts belong to the alpha group of the class Proteobacteria, relating them to Rhizobium and Thiobacillus. Other studies indicate that these subcuticular bacteria may be both abundant within their hosts and widely distributed among the Echinoderms in general.

Symbiodinium Dinoflagellate Endosymbionts in Marine Metazoa and Protists

Dinoflagellate endosymbionts of the genus Symbiodinium, commonly known as zooxanthellae, are found in corals, mollusks (esp. giant clams, the Tridacna), sponges, and foraminifera. These endosymbionts drive the amazing formation of coral reefs by capturing sunlight and providing their hosts with energy for carbonate deposition.

Previously thought to be a single species, molecular phylogenetic evidence over the past couple decades has shown there to be great diversity in Symbiodinium. In some cases there is specificity between host and Symbiodinium clade. More often, however, there is an ecological distribution of Symbiodinium, the symbionts switching between hosts with apparent ease. When reefs become environmentally stressed, this distribution of symbionts is related to the observed pattern of coral bleaching and recovery. Thus the distribution of Symbiodinium on coral reefs and its role in coral bleaching presents one of the most complex and interesting current problems in reef ecology.

Endosymbionts in Protists

Mixotricha paradoxa is a protozoan that lacks mitochondria, however, spherical bacteria live inside the cell and serve the function of the mitochondria. Mixotricha also has three other species of symbionts that live on the surface of the cell.

Paramecium bursaria, a species of ciliate, has a mutualistic symbiotic relationship with green alga called Zoochlorella. The algae live inside the cell, in the cytoplasm.

Bacterial Obligate Endosymbionts in Insects

Scientists classify insect endosymbionts in two broad categories, 'Primary' and 'Secondary'. Primary endosymbionts (sometimes referred to as P-endosymbionts) have been associated with their insect hosts for many millions of years (from 10 to several hundred million years in some cases), they form obligate associations, and display cospeciation with their insect hosts. Secondary endosymbionts exhibit a more recently developed association, are sometimes horizontally trasnferred between hosts, live in the haemolymph of the insects, and are not obligate.

Among primary endosymbionts of insects, the best studied are the pea aphid Acyrthosiphon pisum and its endosymbiont Buchnera sp. APS, and the tsetse fly Glossina morsitans morsitans and its endosymbiont Wigglesworthia glossinidia brevipalpis. As with endosymbiosis in other insects, the symbiosis is obligate in that neither the bacteria nor the insect is viable without the other. Scientists have been unable to cultivate the bacteria in lab conditions outside of the insect. With special nutritionally-enhanced diets, the insects can survive, but are unhealthy, and at best survive only a few generations. These endosymbionts live in specialized insect cells called bacteriocytes (also called mycetocytes), and are maternally-transmitted, i.e. the mother transmits her endosymbionts to her offspring. In some cases, the bacteria are transmitted in the egg, as in Buchnera; in others like Wigglesworthia, they are transmitted via milk to the developing insect embryo.

The primary endosymbionts are thought to help the host by either synthesizing nutrients that the host cannot make itself, or by metabolizing insect waste products into safer forms. For example, the primary role of Buchnera is thought to be to synthesize essential amino acids that the aphid cannot acquire from its natural diet of plant sap. The evidence is (1) when aphids' endosymbionts are killed using antibiotics, they appear healthier when their plant sap diet is supplemented with the appropriate amino acids, and (2) after the Buchnera genome was sequenced, analysis uncovered a large number of genes that likely code for amino acid biosynthesis genes; most bacteria that live inside other organisms do not have such genes, so their existence in Buchnera is noteworthy. Similarly, the primary role of Wigglesworthia is probably to synthesize vitamins that the tsetse fly does not get from the blood that it eats.

Bacteria benefit from the reduced exposure to predators, the ample supply of nutrients and relative environmental stability inside the host.

Genome sequencing reveals that obligate bacterial endosymbionts of insects have among the smallest of known bacterial genomes and have lost many genes that are commonly found in closely related bacteria. Several theories have been put forth to explain the loss of genes. Presumably some of these genes are not needed in the environment of the host insect cell. A complementary theory suggests that the relatively small numbers of bacteria inside each insect decrease the efficiency of natural selection in 'purging' deleterious mutations and small mutations from the population, resulting in a loss of genes over many millions of years. Research in which a parallel phylogeny of bacteria and insects was inferred supports the belief that the primary endosymbionts are transferred only vertically (i.e. from the mother), and not horizontally (i.e. by escaping the host and entering a new host).

Attacking obligate bacterial endosymbionts may present a way to control their insect hosts, many of which are pests or carriers of human disease. For example aphids are crop pests and the tsetse fly carries the organism (trypanosome protozoa) that causes African sleeping sickness. Other motivations for their study is to understand symbiosis, and to understand how bacteria with severely depleted genomes are able to survive, thus improving our knowledge of genetics and molecular biology.

Less is known about secondary endosymbionts. The pea aphid (Acyrthosiphon pisum) is known to contain at least three secondary endosymbionts, Hamiltonella defensa, Regiella insecticola, and

Serratia symbiotica. H. defensa aids in defending the insect from parasitoids.

The chromists and alveolates may be part of a larger group that is ancestrally photosynthetic, called the chromalveolates, but this remains contentious. Otherwise the relationships between the different supergroups are mostly uncertain, and in particular there is dispute about where the root of the evolutionary tree belongs, and as a result what the earliest eukaryotes were like. A few small protist groups have not been related to any of the major supergroups, notably the centrohelids, apusozoans, and ebriids.

Eukaryotes are closely related to archaea, at least in terms of nuclear DNA and genetic machinery. In other respects, such as membrane composition, they are similar to eubacteria. Three main explanations for this have been proposed:

- Eukaryotes resulted from the complete fusion of two or more cells, the cytoplasm forming from a eubacterium and the nucleus from an archaeon (alternatively a virus).
- Eukaryotes developed from Archaea, and acquired their eubacterial characteristics from the proto-mitochondrion.
- Eukaryotes and Archaea developed separately from a modified eubacterium.

The origin of the endomembrane system and mitochondria are also disputed. The phagotrophic hypothesis states the membranes originated with the development of endocytosis and later specialized; mitochondria were acquired by ingestion, like plastids. The syntrophic hypothesis states that the proto-eukaryote relied on the proto-mitochondrion for food, and so ultimately grew to surround it; the membranes originate later, in part thanks to mitochondrial genes (the hydrogen hypothesis is one particular version).

A new hypothetical theory thinks of the Eucarya ancestor as the Urkaryote. This ancestor is only the precursor of the Eucarya and is not the ancestor of the Archaea or Bacteria. The common ancestor of all three of these domains is a more primitive organism with a variety of different cell structures (less vs more complex structures, pseudomurein vs peptidoglycan cell wall, rRNA differences, lipid bilayer membrane vs phospholipid bilayer membrane, nucleoid region vs defined nucleus, differences in metabolic processes, Enther Dnodroff Pathway vs Glycolysis Pathway, etc).

Molecular studies place them in a supergroup called the opisthokonts, which also include the fungi and a few small parasitic protists. The name comes from the posterior location of the flagellum in motile cells, such as most animal spermatozoa, whereas other eukaryotes tend to have anterior flagella.

The first fossils that might represent animals appear towards the end of the Precambrian, around 575 million years ago, and are known as the Ediacaran or Vendian biota. These are difficult to relate to later fossils, however. Some may represent precursors of modern phyla, but they may be separate groups, and it is possible they are not really animals at all. Aside from them, most known animal phyla make a more or less simultaneous appearance during the Cambrian period, about 542 million years ago. It is still disputed whether this event, called the Cambrian explosion, represents a rapid divergence between different groups or a change in conditions that made fossilization possible.

FUNGUS

The fungi (singular fungus) are a kingdom of eukaryotic organisms. They are heterotrophic and digest their food externally, absorbing nutrient molecules into their cells. Yeasts, molds, and mushrooms are examples of fungi. The branch of biology involving the study of fungi is known as mycology.

Fungi often have important symbiotic relationships with other organisms. Mycorrhizal symbiosis between plants and fungi is particularly important; over 90% of all plant species engage in some kind of mycorrhizal relationship with fungi and are dependent upon this relationship for survival. Fungi are also used extensively by humans: yeasts are responsible for fermentation of beer and bread, and mushroom farming and gathering is a large industry in many countries.

Fungi and bacteria are the primary decomposers of organic matter in most terrestrial ecosystems.

Phylogeny and Classification of Fungi

Fungi were originally classified as plants, however they have since been separated as they are heterotrophs. This means they do not fix their own carbon through photosynthesis, but use carbon fixed by other organisms for metabolism. Fungi are now thought to be more closely related to animals than to plants, and are placed with animals in the monophyletic group of opisthokonts. For much of the Paleozoic Era, the fungi appear to have been aquatic. The first land fungi probably appeared in the Silurian, right after the first land plants appeared, even though their fossils are fragmentary. Fungi absorb their food while animals ingest it; also unlike animals, the cells of fungi have cell walls. For these reasons, these organisms are placed in their own kingdom, Fungi, or Eumycota. The Fungi are a monophyletic group, meaning all varieties of fungi come from a common ancestor. The monophyly of the fungi has been confirmed through repeated tests of molecular phylogenetics; shared ancestral traits include chitinous cell walls and heterotrophy by absorption, along with other shared characteristics.

The taxonomy of the Fungi is in a state of rapid flux at present, especially due to recent papers based on DNA comparisons, which often overturn the assumptions of the older systems of classification. There is no unique generally accepted system at the higher taxonomic levels and there are constant name changes at every level, from species upwards. Web sites such as Index Fungorum, ITIS and Wikispecies define preferred up-to-date names (with cross-references to older synonyms), but do not always agree with each other or with names in Wikipedia in its various language variants.

Types of Fungi

The major divisions (phyla) of fungi are mainly classified based on their sexual reproductive structures. Currently, five divisions are recognized:

- The Chytridiomycota are commonly known as chytrids. These fungi produce zoospores that are capable of moving on their own through liquid menstrua by simple flagella.
- The Zygomycota are known as zygomycetes and reproduce sexually with meiospores called zygospores and asexually with sporangiospores. Black bread mold (Rhizopus stolonifer) is a common species that belongs to this group; another is Pilobolus, which shoots specialized structures through the air for several meters. Medically relevant genera include Mucor, Rhizomucor, and Rhizopus. Molecular phylogenetic investigation has shown the zygomycota to be a polyphyletic group.
- Members of the Glomeromycota are also known as the arbuscular mycorrhizal fungi. Only one species has been observed forming zygospores; all other species only reproduce asexually. This is an ancient association, with evidence dating to 350 million years ago.
- The Ascomycota, commonly known as sac fungi or ascomycetes, form meiotic spores called ascospores, which are enclosed in a special sac-like structure called an ascus. This division includes morels, some mushrooms and truffles, as well as single-celled yeasts and many species that have only been observed undergoing asexual reproduction. Because the products of meiosis are retained within the sac-like ascus, several ascomyctes have been used for elucidating principles of genetics and heredity (e.g. Neurospora crassa).
- Members of the Basidiomycota, commonly known as the club fungi or basidiomycetes, produce meiospores called basidiospores on club-like stalks called basidia. Most common mushrooms belong to this group, as well as rust (fungus) and smut fungi, which are major pathogens of grains.

Although the water moulds and slime moulds have traditionally been placed in the kingdom Fungi and those who study them are still called mycologists, they are not true fungi.

Unlike true fungi, the water moulds and slime moulds do not have cell walls made of chitin. In the 5-kingdom system, they are currently placed in the kingdom Protista. Water moulds are descended from algae, and are placed within the phylum Oomycota, within the Kingdom Protista.

Reproduction

Fungi may reproduce sexually or asexually. In asexual reproduction, the offspring are genetically identical to the "parent" organism (they are clones). During sexual reproduction, a mixing of genetic material occurs so that the offspring exhibit traits of both parents. Many species can use both strategies at different times, while others are apparently strictly sexual or strictly asexual. Sexual reproduction has not been observed in some fungi of the Glomeromycota and Ascomycota. These are commonly referred to as Fungi imperfecti or Deuteromycota.

Yeasts and other unicellular fungi can reproduce simply by budding, or "pinching off" a new cell. Many multicellular species produce a variety of different asexual spores that are easily dispersed and resistant to harsh environmental conditions. When the conditions are right, these spores will germinate and colonize new habitats.

Sexual reproduction in fungi is somewhat different from that of animals or plants, and each fungal division reproduces using different strategies. Fungi that are known to reproduce sexually all have a haploid stage and a diploid stage in their life cycles. Ascomycetes and basidiomycetes also go through a dikaryotic stage, in which the nuclei inherited by the two parents do not fuse right away, but remain separate in the hyphal cells.

In zygomycetes, the haploid hyphae of two compatible individuals fuse, forming a zygote, which becomes a resistant zygospore. When this zygospore germinates, it quickly undergoes meiosis, generating new haploid hyphae and asexual sporangiospores. These sporangiospores may then be distributed and germinate into new genetically-identical individuals, each producing their own haploid hyphae. When the hyphae of two compatible individuals come into contact with one another, they will fuse and generate new zygospores, thus completing the cycle.

In ascomycetes, when compatible haploid hyphae fuse with one another, their nuclei do not immediately fuse. The dikaryotic hyphae form structures called asci (sing. ascus), in which karyogamy (nuclear fusion) occurs. These asci are embedded in an ascocarp, or fruiting body, of the fungus. Karyogamy in the asci is followed immediately by meiosis and the production of ascospores. The ascospores are disseminated and germinate to form new haploid mycelium. Asexual conidia may be produced by the haploid mycelium. Many ascomycetes appear to have lost the ability to reproduce sexually and reproduce only via conidia.

Sexual reproduction in basidiomycetes is similar to that of ascomycetes. Sexually compatible haploid hyphae fuse to produce a dikaryotic mycelium. This leads to the production of a basidiocarp. The most commonly-known basidiocarps are mushrooms, but they may also take many other forms. Club-like structures known as basidia generate haploid basidiospores following karyogamy and meiosis. These basidiospores then germinate to produce new haploid mycelia.

Ecological Role

Although often inconspicuous, fungi occur in every environment on Earth and play very important roles in most ecosystems. Along with bacteria, fungi are the major decomposers in most terrestrial (and some aquatic) ecosystems, and therefore play a critical role in biogeochemical cycles and in many food webs.

Many fungi are important as partners in symbiotic relationships with other organisms, as mutualists, parasites, or commensalists, as well as in symbiotic relationships that do not fall neatly into any of these categories. One of the most important of these relationships are various types of mycorrhiza, which is a kind of mutualistic relationship between fungi and plants, in which the plant's roots are closely associated with fungal hyphae and other structures. The plant donates to the fungus sugars and other carbohydrates that it manufactures from photosynthesis, while the fungus donates water and mineral nutrients that the hyphal network is able to find much more efficiently than the plant roots alone can, particularly phosphorus. The fungi also protect against diseases and pathogens and provide other benefits to the plant.

Recently, plants have been found to use mycorrhizas to deliver carbohydrates and other nutrients to other plants in the same community and in some cases can make plant species that would normally exclude each other able to coexist in the same plant community. Such mycorrhizal communities are called "common mycorrhizal networks". Over 90% of the plant species on Earth are dependent on mycorrhizae of one type or another in order to survive, and it is hypothesized that the presence of terrestrial fungi may have been necessary in order for the first plants to colonize land.

Lichens are formed by a symbiotic relationship between algae or cyanobacteria (referred to in lichens as "photobionts") and fungi (mostly ascomycetes of various kinds and a few basidiomycetes), in which individual photobiont cells are embedded in a complex of fungal tissue. As in mycorrhizas, the photobiont provides sugars and other carbohydrates while the fungus provides minerals and water. The functions of both symbiotic organisms are so closely intertwined that they function almost as a single organism.

Certain insects also engage in mutualistic relationships with various types of fungi. Several groups of ants cultivate various fungi in the Agaricales as their primary food source, while ambrosia beetles cultivate various kinds of fungi in the bark of trees that they infest.

Some fungi are parasites on plants, animals (including humans), and even other fungi. Pathogenic fungi are responsible for numerous diseases, such as athlete's foot and ringworm in humans and Dutch elm disease in plants. Some fungi are predators of nematodes, which they capture using an array of devices such as constricting rings or adhesive nets.

Human uses of Fungi

Fungi have a long history of use by humans. Many types of mushrooms and other fungi are eaten, including button mushrooms, shiitake mushrooms, and oyster mushrooms. Of course, many species of mushrooms are poisonous and are responsible for numerous cases of sickness and death every year. A type of single-celled fungus called yeast is used in baking bread and fermenting alcoholic beverages, while mycelial fungus is used to make Shoyu (soy sauce) and tempeh. Fungi are also used to produce industrial chemicals like lactic acid, antibiotics and even to make stonewashed jeans. Some types of fungi are ingested for their psychedelic properties, both recreationally and religiously.

Edible and Poisonous Fungi

Some of the most well-known types of fungi are the edible and poisonous mushrooms. Many species are commercially raised, but others must be harvested from the wild. Button mushrooms (Agaricus bisporus) are the most commonly eaten species, used in salads, soups, and many other dishes. Portobello mushrooms are the same species, but are allowed to grow to a much larger size. Other commercially-grown mushrooms that have gained in popularity in the West and are often available fresh in grocery stores include straw mushrooms (Volvariella volvacea), oyster mushrooms (Pleurotus ostreatus), shiitakes (Lentinula edodes), and enokitake (Flammulina spp.).

There are many more mushroom species that are harvested from the wild for personal consumption or commercial sale. Milk mushrooms, morels, chanterelles, truffles, black trumpets, and porcini mushrooms (also known as king boletes) all command a high price on the market. They are often used in gourmet dishes.

It is also a common practice to permit the growth of specific species of mold in certain types of cheeses that give them their unique flavor. This mold is non-toxic and is safe for human consumption. This accounts for the blue colour in cheeses such as Roquefort or Stilton. Hundreds of mushroom species are toxic to humans, causing anything from upset stomachs to hallucinations to death. Some of the most deadly belong to the genus Amanita, including A. virosa (the "destroying angel") and A. phalloides (the "death cap"). Stomach cramps, vomiting, and diarrhea usually occur within 6-24 hours after ingestion of these mushrooms, followed by a brief period of remission (usually 1-2 days). Patients often fail to present themselves for treatment at this time, assuming that they have recovered.

However, within 2-4 weeks liver and kidney failure leads to death if untreated. There is no antidote for the toxins in these mushrooms, but kidney dialysis and administration of corticosteroids may help. In severe cases, a liver transplant may be necessary (Kaminstein 2002). It is difficult to identify a "safe" mushroom without proper training and knowledge, thus it is often advised to assume that a mushroom in the wild is poisonous and leave it alone.

Fly agaric mushrooms (A. muscaria) are also responsible for a large number of poisonings, but these cases rarely result in death. The most common symptoms are nausea and vomiting, drowsiness, and hallucinations. In fact, this species is used ritually and recreationally for its hallucinogenic properties. However, if it is taken in over a long period of time (regularly over more than six months), this species might cause a temporary loss of sight, which can last from several minutes to an hour.

Fungi in the Biological Control of Pests

Many fungi compete with other organisms, or directly infect them. Some of these fungi are considered beneficial because they can restrict, and sometimes eliminate, the populations of noxious organisms like pest insects, mites, weeds, nematodes and other fungi, such as those that kill plants. There is much interest on the manipulation of these beneficial fungi for the biological control of pests. Some of these fungi can be used as biopesticides, like the ones that kill insects (entomopathogenic fungi). Specific examples of fungi that have been developed as bioinsecticides are Beauveria bassiana, Metarhizium anisopliae, Hirsutella, Paecilomyces fumosoroseus, and Verticillium lecanii (=Lecanicillium lecanii).

POSSIBLE BASIS OF EXTRATERRESTRIAL LIFE

Biochemistry

All life on Earth is based on the building block element carbon with water as the solvent in which bio-chemical reactions take place. The combination of carbon and water in the chemical form (CH<sb>2O)n, is the chemical form of the sugars, which as well as providing the energy on which life depends (largely through the oxidation of glucose, a six carbon sugar), also provides structural elements for life (such as the sugar ribose, a five carbon sugar, in the molecules DNA and RNA). Life requires carbon in both reduced (methane derivatives) and partially-oxidized (carbon oxides) states. It also requires nitrogen as a reduced ammonia derivative in all proteins, sulfur as a derivative of hydrogen sulfide in some necessary proteins, and phosphorus oxidized to phosphates in genetic material and in energy transfer. Adequate water as a solvent supplies adequate oxygen as constituents of biochemical substances.

Pure water is useful because it has a neutral pH, due to its continued dissociation between hydroxide and hydronium ions.

As a result, it can dissolve both positive metallic ions and negative non-metallic ions with equal ability. Furthermore, the fact that organic molecules can be either hydrophobic (repelled by water) or hydrophilic (soluble in water) creates the ability of organic compounds to orient themselves to form water-enclosing membranes. The fact that solid water (ice) is less dense than liquid water also means that ice floats, thereby preventing Earth's oceans from slowly freezing solid. Additionally, the Van der Waals forces between water molecules gives it an ability to store energy with evaporation, which upon condensation is released. This helps moderate climate, cooling the tropics and warming the poles, helping to maintain a thermodynamic stability needed for life.

- Carbon is fundamental to terrestrial life for its immense flexibility in creating covalent chemical bonds with a variety of non-metallic elements, principally nitrogen, oxygen and hydrogen. Carbon dioxide and water together enable the storage of solar energy in sugars, such as glucose. The oxidation of glucose releases biochemical energy needed to fuel all other biochemical reactions

$$6CO_2 + 6H_2O + \text{sun energy} \rightarrow C_6H_{12}O_6 + 6O_2$$

$$C_6H_{12}O_6 + 6O_2 \rightarrow 6CO_2 + 6H_2O + \text{biochemical energy}$$

The ability to form organic acids (-COOH) and amine bases (NH2-) gives it the possibility of neutralisation dehydrating reactions to build long polymer peptides and catalytic proteins from monomer amino acids, and with phosphates to build not only DNA, the information storing molecule of inheritance, but also adenosine triphosphate (ATP) the principle energy "currency" of cellular life.

Given their relative abundance and usefulness in sustaining life it has long been assumed that life forms elsewhere in the universe will also utilize these basic components. However, other elements and solvents might be capable of providing a basis for life. Silicon is usually considered the most likely alternative to carbon, though this remains improbable. Silicon life forms are proposed to have a crystalline morphology, and are theorized to be able to exist in high temperatures, such as planets closer to the sun. Life forms based in ammonia rather than water are also considered, though this solution appears less optimal than water.

Indeed, technically life is little more than any self-replicating reaction, which could arise in a great many conditions and with various ingredients, though carbon-oxygen within the liquid temperature range of water seems most conducive. Suggestions have even been made that self-replicating reactions of some sort could ocur within the plasma of a star, though it would be highly unconventional, since plasma is essentially the fourth state of matter, where electrons are not bound in their orbits around atomic nuclei.

Evolution and Morphology

Along with the biochemical basis of extraterrestrial life, there remains a broader consideration of evolution and morphology. Science fiction has long shown a bias towards humanoid and/or reptilian forms. The classical alien is light green or grey skinned, with a large head, and the typical four limb and two to five digit structure-i.e., it is fundamentally humanoid with a large brain to indicate great intelligence. Other subjects from animal mythos such as felines and insects have also featured strongly in fictional representations of aliens.

A division has been suggested between universal and parochial (narrowly restricted) characteristics. Universals are features which have evolved independently more than once on Earth (and thus presumably are not difficult to develop) and are so intrinsically useful that species will inevitably tend towards them. These include flight, sight, photosynthesis and limbs, all of which have evolved several times here on Earth with differing materialization. There are a huge variety of eyes, for example, many of which have radically different working schematics as well as different visual foci: the visual spectrum, infrared, polarity and echolocation. Parochials, by contrast, are essentially arbitrary evolutionary forms which often serve little inherent utility (or at least have a function which can be equally served by dissimilar morphology) and probably will not be replicated. Parochials include the five digits of mammals and the curious and often fatal conjunction of the feeding and breathing passages found within many animals, although it is possible this conjunction allowed for the evolution of human speech.

A consideration of which features are ultimately parochial, challenges many taken-for-granted notions about morphological necessity. Skeletons, which are essential to large terrestrial organisms according to the experts of the field of Gravitational biology, are almost assuredly to be replicated elsewhere in one form or another, yet the vertebrate spine-while a profound development on Earth-is just as likely to be unique. Similarly, it is reasonable to expect some type of egg laying amongst off-Earth creatures but the mammary glands which set apart mammals might be a singular case.

The assumption of radical diversity amongst putative extraterrestrials is by no means settled. While many exobiologists do stress that the enormously heterogeneous nature of Earth life foregrounds even greater variety in space, others point out that convergent evolution may dictate substantial similarities between Earth and off-Earth life. These two schools of thought are called "divergionism" and "convergionism", respectively.

Beliefs in Extraterrestrial Life

Ancient and Early Modern Ideas

Belief in extraterrestrial life may have been present in ancient Egypt, China, Babylon, India and Sumer, although in these societies, cosmology was fundamentally supernatural and the notion of alien life is difficult to distinguish from that of gods, demons, and such. The first important Western thinkers to argue systematically for a universe full of other planets and, therefore, possible extraterrestrial life were the ancient Greek writers Thales and his student Anaximander in the 7th and 6th centuries B.C. The atomists of Greece took up the idea, arguing that an infinite universe ought to have an infinity of populated worlds. Ancient Greek cosmology worked against the idea of extraterrestrial life in one critical respect, however: the geocentric universe, championed by Aristotle and codified by Ptolemy, favored the Earth and Earth-life (Aristotle denied there could be a plurality of worlds) and seemingly rendered extraterrestrial life philiosophically untenable. Lucian in his novels described inhabitants of the Moon and other celestial bodies as humanoids, but with significant differences from humans.

Authors of ancient Jewish sources also considered extraterrestrial life. The Talmud states that there are at least 18,000 other worlds, but provides little elaboration on the nature of the worlds and on whether they are physical or spiritual. Based on this, however, the medieval exposition "Sefer HaB'rit" posits that extraterrestrial creatures exist but that they have no free will (and are thus equivalent to animal life). It adds that human beings should not expect creatures from another world to resemble earthly life, any more than sea creatures resemble land animals.

Hindu beliefs of endlessly repeated cycles of life have lead to descriptions of multiple worlds in existence and their mutual contacts (Sanskrit word Sampark means 'contact' as in Mahasamparka = the great contact). However the relevance of such descriptions have to be evaluated in the context of understanding of geography and science at those times. Within Islam, the statement of the Qur'an "All praise belongs to God, Lord of all the worlds" indicates multiple universal bodies and maybe even multiple universes that may indicate extraterrestrial and even extradimensional life. Surat Al-Jinn also mentioned a statement from a Jinn regarding the current status and ability of his group in the heavens.

When Christianity spread throughout the West, the Ptolemaic system became very widely accepted, and although the Church never issued any formal pronouncement on the question of alien life at least tacitly the idea was aberrant. In 1277 the Bishop of Paris, Étienne Tempier, did overturn Aristotle on one point: God could have created more than one world (given His omnipotence) yet we know by revelation He only made one. Taking a further step and arguing that aliens actually existed remained rare. Notably, Cardinal Nicholas of Cusa speculated about aliens on the moon and sun.

This situation changed, however, with the dramatic shift in thinking initiated by the invention of the telescope and the Copernican assault on geocentric cosmology. Once it became clear that the Earth was merely one planet amongst countless bodies in the universe the extraterrestrial idea moved towards the scientific mainstream. God's omnipotence, it could be argued, not only allowed for other worlds and other life, on some level it necessitated them. The best known early-modern proponent of such ideas was Giordano Bruno, who argued in the 16th century for an infinite universe in which every star is surrounded by its own solar system; he was eventually burned at the stake by the Catholic church for his heretical ideas. In the early 17th century the Czech astronomer Anton Maria Schyrleus of Rheita mused that "if Jupiter has..inhabitants..they must be larger and more beautiful than the inhabitants of the Earth, in proportion to the [characteristics] of the two spheres." Dominican monk Tommaso Campanella wrote about a Solarian alien race in his Civitas Solis.

Such comparisons also appeared in poetry of the era. In "The Creation: a Philosophical Poem in Seven Books" (1712) Sir Richard Blackmore observed: "We may pronounce each orb sustains a race / Of living things adapted to the place". The didactic poet Henry More took up the classical theme of the Greek Democritus in "Democritus Platonissans, or an Essay Upon the Infinity of Worlds" (1647). With the new relative viewpoint that the Copernican revolution had wrought, he suggested "our world's sunne / Becomes a starre elsewhere." Fontanelle's "Conversations on the Plurality of Worlds" (translated into English in 1686) offered similar excursions on the possibility of extraterrestrial life, expanding rather than denying the creative sphere of a Maker.

The possibility of extraterrestrials remained a widespread speculation as scientific discovery accelerated. William Herschel, the discoverer of Uranus, was one of many 18th-19th century astronomers convinced that our Solar System, and perhaps others, would be well populated by alien life. Other luminaries of the period who championed "cosmic pluralism" included Immanuel Kant and Benjamin Franklin. At the height of the Enlightenment even the Sun and Moon were considered candidates for hosting aliens.

Extraterrestrials and the Modern Era

This enthusiasm towards the possibility of alien life continued well into the 20th century. Indeed, the roughly three centuries from the Scientific Revolution through the beginning of the modern era of solar system probes were essentially the zenith for belief in extraterrestrials in the West: many astronomers and other secular thinkers, at least some religious thinkers, and much of the general public were largely satisfied that aliens were a reality. This trend was finally tempered as actual probes visited potential alien abodes in the solar system. The moon was decisively ruled out as a possibility, while Venus and Mars, long the two main candidates for extraterrestrials, showed no obvious evidence of current life.

The other large moons of our system which have been visited appear similarly lifeless, though the interesting geothermic forces observed (Io's volcanism, Europa's ocean, Titan's thick atmosphere)

have underscored how broad the range of potentially habitable environments may be. Although the hypothesis of a deliberate cosmic silence of advanced extraterrestrials is also a possibility, the failure of the SETI program to detect anything resembling an intelligent radio signal after four decades of effort has partially dimmed the optimism that prevailed at the beginning of the space age. Emboldened critics view the search for extraterrestrials as unscientific, despite the fact the SETI program is not the result of a continuous, dedicated search but instead utilizes what resources and manpower it can, when it can.

Thus, the three decades preceding the turn of the second millennium saw a crossroads reached in beliefs in alien life. The prospect of ubiquitous, intelligent, space-faring civilizations in our solar system appears increasingly dubious to many scientists. Still, in the words of SETI's Frank Drake, "All we know for sure is that the sky is not littered with powerful microwave transmitters." Drake has also noted that it is entirely possible advanced technology results in communication being carried out in some way other than conventional radio transmission. At the same time, the data returned by space probes and giant strides in detection methods have allowed science to begin delineating habitability criteria on other worlds and to confirm that, at least, other planets are plentiful though aliens remain a question mark.

At present, some enthusiasts in the topic believe that extraterrestrial beings regularly visit or have visited the Earth. Some think that unidentified flying objects observed in the skies are in fact sightings of the spacecraft of intelligent extraterrestrials, and even claim to have met such beings. Crop circle patterns have also been attributed to the actions of extraterrestrials by some believers, although many were later found to be hoaxes. While one recent scientific paper published in the peer-reviewed Journal of the British Interplanetary Society has urged a re-evaluation of the UFO phenomenon, as of this time mainstream scientific opinion holds that such claims are unsupportable by the evidence currently available and unlikely to be true. The possible existence of primitive (microbial) life outside of Earth is much less controversial to mainstream scientists although at present no direct evidence of such life has been found. Indirect evidence has been offered for the current existence of primitive life on the planet Mars; however, the conclusions that should be drawn from such evidence remain in debate.

Scientific Search for Extraterrestrial Life

The scientific search for extraterrestrial life is being carried out in two different ways, directly and indirectly.

Direct Search

Scientists are directly searching for evidence of unicellular life within the solar system, carrying out studies on the surface of Mars and examining meteors that have fallen to Earth. A mission is also proposed to Europa, one of Jupiter's moons with a possible liquid water layer under its surface, which might contain life.

There is some limited evidence that microbial life might possibly exist or have existed on Mars. An experiment on the Viking Mars lander reported gas emissions from heated Martian soil that some argue are consistent with the presence of microbes. However, the lack of corroborating evidence from other experiments on the Viking indicates that a non-biological reaction is a more likely hypothesis. Recently, Circadian rhythms have been allegedly discovered in Viking data. The interpretation is controversial. Independently in 1996 structures resembling bacteria were reportedly discovered in a meteorite, ALH84001, thought to be formed of rock ejected from Mars. This report is also controversial and scientific debate continues.

In February 2005, NASA scientists reported that they had found strong evidence of present life on Mars. The two scientists, Carol Stoker and Larry Lemke of NASA's Ames Research Center, based their claims on methane signatures found in Mars' atmosphere that resemble the methane production of some forms of primitive life on Earth, as well as their own study of primitive life near the Rio Tinto river in Spain. NASA officials soon denied the scientists' claims, and Stoker herself backed off from her initial assertions. Though such findings are still very much in debate, support among scientists for the belief in the existence of life on

Mars seems to be growing. In an informal survey conducted at the conference in which the European Space Agency presented its findings, 75 percent of the scientists in attendance reported to believe that life once existed on Mars; 25 percent reported a belief that life currently exists there.

Indirect Search

It is theorised that any technological society in space will be transmitting information. Projects such as SETI are conducting an astronomical search for radio activity that would confirm the presence of intelligent life. A related suggestion is that aliens might broadcast pulsed and continuous laser signals in the optical as well as infrared spectrum; laser signals have the advantage of not "smearing" in the interstellar medium and may prove more conducive to communication between the stars.

Extrasolar Planets

Astronomers also search for extrasolar planets that would be conducive to life, especially those like OGLE-2005-BLG-390Lb which have been found to have Earth-like qualities. Current radiodetection methods have been inadequate for such a search, as the resolution afforded by recent technology is inadequate for detailed study of extrasolar planetary objects. Future telescopes should be able to image planets around nearby stars, which may reveal the presence of life (either directly or through spectrography which would reveal key information such as the presence of free oxygen in a planet's atmosphere). Darwin is an ESA mission designed to find Earth-like planets, and analyse their atmosphere. It has been argued that one of the best candidates for the discovery of life-supporting planets may be Alpha Centauri, the closest star system to Earth, given that two of the three stars in the system are broadly sun-like.

Extraterrestrial Life in the Solar System

Many bodies in the Solar System have been suggested as being capable of containing conventional organic life. The most commonly suggested ones are listed below; of these, four of the seven are moons, and are thought to have large bodies of underground liquid, where life may have evolved in a similar fashion to deep sea vents.

- Mars-Liquid water is widely thought to have existed on Mars in the past and there may still be liquid water beneath the surface. Methane was found in the atmosphere of Mars. Recent photographs from Mars Global Surveyor show evidence of recent (within 10 years) flows of a liquid on the Red Planet's frigid surface. There is however uncertainty as to whether this was liquid water or CO_2 Main article: Life on Mars
- Europa-Europa may contain liquid water beneath its 100-mile ice layer, vents on the bottom of the ocean warm the ice so that 60 miles of liquid could exist beneath the ice layer, perhaps capable of supporting microbes and simple plants
- Jupiter-Possible supporter of floating animals, as hypothesized by Carl Sagan. This point of view is somewhat controversial due to the fact that these creatures would not be water-based, but ammonia-based.
- Ganymede-Possible underground ocean.
- Callisto-Possible underground ocean.
- Saturn-Possible floating creatures.
- Titan (Saturn's largest moon)-The only known moon with a significant atmosphere was recently visited by the Huygens probe. Latest discoveries indicate that there is no global or widespread ocean, but small and/or seasonal liquid hydrocarbon lakes are almost surely present on the surface.

Numerous other bodies have been suggested as potential hosts for microbial life. Fred Hoyle has proposed that life might exist on comets, as some Earth microbes managed to survive on a lunar probe for many years. However, it is considered highly unlikely that complex multicellular organisms of the conventional chemistry of terrestrial life (animals, plants) could exist under these living conditions.

8

THE HUMAN GENOME

GENERAL ORGANIZATION OF THE HUMAN GENOME

The human genome is the term used to describe the total genetic information (DNA content) in human cells. It really comprises two genomes: a complex nuclear genome which accounts for 99.9995% of the total genetic information, and a simple mitochondrial genome which accounts for the remaining 0.0005%.

The nuclear genome provides the great bulk of essential genetic information, most of which specifies polypeptide synthesis on cytoplasmic ribosomes.

Mitochondria possess their own ribosomes and the few polypeptide-encoding genes in the mitochondrial genome produce mRNAs which are translated on the mitochondrial ribosomes. However, the mitochondrial genome specifies only a very small portion of the specific mitochondrial functions; the bulk of the mitochondrial polypeptides are encoded by nuclear genes and are synthesized on cytoplasmic ribosomes, before being imported into the mitochondria.

Like other complex genomes, a sizeable component of the human genome is made up of noncoding DNA. In addition, the human genome is representative of mammalian genomes and other complex genomes in having a considerable amount of repetitive DNA, including both noncoding repetitive DNA and multiple copy genes and gene fragments.

The mitochondrial genome consists of a small circular DNA duplex which is densely packed with genetic information

General structure and inheritance of the mitochondrial genome: The human mitochondrial genome is defined by a single type of circular double-stranded DNA whose complete nucleotide sequence has been established (Anderson *et al.*, 1981). It is 16 569 bp in length and is 44% (G + C). The two DNA strands have significantly different base compositions: the heavy (H) strand is rich in guanines, the light (L) strand is rich in cytosines. Although the mitochondrial DNA is principally double-stranded, a small section is defined by a triple DNA strand structure. This is because a short segment of the heavy strand is replicated for a second time, giving a structure known as 7S DNA. Human cells usually contain thousands of copies of the double-stranded mitochondrial DNA molecule. Accordingly, although a single mitochondrial DNA duplex has only about 1/8000 as much DNA as an average sized chromosome, the total mitochondrial DNA complement can account for up to about 0.5% of the DNA in a nucleated somatic cell.

During zygote formation, a sperm cell contributes its nuclear genome, but not its mitochondrial genome, to the egg cell. Consequently, the mitochondrial genome of the zygote is determined exclusively by that originally found in the unfertilized egg. The mitochondrial genome is therefore maternally inherited: males and females both inherit their mitochondria from their mother but males cannot transmit their mitochondria to subsequent generations.

Mitochondrial genes: The human mitochondrial genome contains 37 genes: 28 are encoded by the heavy strand, and nine by the light strand. Of the 37 genes, a total of 24 specify a mature RNA product: 22 mitochondrial tRNA molecules and two mitochondrial rRNA molecules, a 23S rRNA (a component of the large subunit of mitochondrial ribosomes) and a 16S rRNA (a component of the small subunit of the mitochondrial ribosomes). The remaining 13 genes encode polypeptides which are synthesized on mitochondrial ribosomes.

Each of the 13 polypeptides encoded by the mitochondrial genome is a subunit of one of the mitochondrial respiratory complexes, the multichain enzymes of oxidative phosphorylation which are engaged in the production of ATP. Note, however, that there are a total of about 100 different polypeptide subunits in the mitochondrial oxidative phosphorylation system, and so the vast majority are encoded by nuclear genes. All other mitochondrial proteins, including numerous enzymes, transport proteins, structural proteins etc., are encoded by the nuclear genome and are translated on cytoplasmic ribosomes before being imported into the mitochondria.

The mitochondrial genetic code: The mitochondrial genetic code (which is used to decipher only 13 different mitochondrial mRNAs on mitochondrial ribosomes) differs slightly from the nuclear genetic code (which specifies perhaps about 70 000–80 000 different mRNAs on cytoplasmic ribosomes). The mitochondrial genome encodes all the ribosomal RNA and tRNA molecules it needs for synthesizing proteins but relies on nuclear-encoded genes to provide all other components (such as the protein components of mitochondrial ribosomes, amino acyl tRNA synthetases, etc.).

As there are only 22 different types of human mitochondrial tRNA, individual tRNA molecules need to be able to interpret several different codons. Eight of the 22 tRNA molecules have anticodons which are each able to recognize families of four codons differing only at the third base, and 14 recognize pairs of codons which are identical at the first two base positions and share either a purine or a pyrimidine at the third base. Between them, therefore, the 22 mitochondrial tRNA molecules can recognize a total of 60 codons [(8 × 4) + (14 × 2)]. The remaining four codons, UAG, UAA, AGA and AGG cannot be recognized by mitochondrial tRNA and act as stop codons.

In addition to their differences in genetic capacity and different genetic codes, the mitochondrial and nuclear genomes differ in many other aspects of their organization and expression.

Coding and noncoding DNA: Unlike its nuclear counterpart, the human mitochondrial genome is extremely compact:

approximately 93% of the DNA sequence represents coding sequence. All 37 mitochondrial genes lack introns and they are tightly packed (on average one per 0.45 kb). The coding sequences of some genes (notably those encoding the sixth and eighth subunits of the mitochondrial ATPase) show some overlap and, in most other cases, the coding sequences of neighboring genes are contiguous or separated by one or two noncoding bases. Some genes even lack termination codons; to overcome this deficiency, UAA codons have to be introduced at the post-transcriptional level.

The only significant region lacking any known coding DNA is the displacement (D) loop region. This is the region in which a triple-stranded DNA structure is generated by synthesizing an additional short piece of the H-strand DNA, known as 7S DNA. The replication of both the H and L strands is unidirectional and starts at specific origins. In the former case, the origin is in the D loop and only after about two-thirds of the daughter H strand has been synthesized (by using the L strand as a template and displacing the old H strand) does the origin for L strand replication become exposed. Thereafter, replication of the L strand proceeds in the opposite direction, using the H strand as a template. The D loop also contains the predominant promoter for transcription of both the H and L strands. Unlike transcription of nuclear genes, in which individual genes are almost always transcribed separately using individual promoters, transcription of the mitochondrial DNA starts from the promoters in the D loop region and continues, in opposing directions for the two different strands, round the circle to generate large multigenic transcripts. The mature RNAs are subsequently generated by cleavage of the multigenic transcripts.

The nuclear genome is distributed between 24 different types of DNA duplex which show considerable regional variation in base composition and gene density

Size and banding patterns of human chromosomes: The nucleus of a human cell contains more than 99% of the cellular DNA. The nuclear genome is distributed between 24 different types of linear double-stranded DNA molecule, each of which has histones and other nonhistone proteins bound to it, constituting a chromosome. The 24 different chromosomes (22 types of autosome and two sex chromosomes, X and Y) can easily be differentiated by chromosome banding techniques, and have been classified into groups largely according to size and, to some extent, centromere position. In addition to the primary constriction (centromere) present on each chromosome, the long arms of chromosomes 1, 9 and 16 possess so-called secondary constrictions (light staining, apparently uncoiled chromosomal regions) which, like the centromeres, are composed of constitutive heterochromatin. By comparison with the size of a mitochondrial DNA molecule, an average size human chromosome has an enormous amount of DNA, approximately 130 Mb on average, but varying between approximately 50 and 260 Mb. In a 550 band metaphase chromosome preparation, an average band corresponds to about 6 Mb of DNA.

Base composition in the human nuclear genome: Since the entire nucleotide sequence of the human mitochondrial genome is known, its precise base composition is known. The sequence of the human nuclear genome is still being established (and is not expected to be finished before 2003), but current estimates suggest a figure of about 42% GC. However, the proportion of specific combinations of nucleotides can vary considerably. Like other vertebrate nuclear genomes, for example, the human nuclear genome has a conspicuous shortage of the dinucleotide CpG (that is, neighboring cytosine and guanine residues on the same DNA strand in the 5' → 3' direction). Taking the average figure of 42% GC, the individual base frequencies are : C = G = 0.21, and so the expected frequency for the dinucleotide CpG is $(0.21)^2 = 0.0441$. However, the observed frequency of the CpG dinucleotide is approximately one-fifth of this.

In vertebrate DNA, cytosine residues occurring in CpG dinucleotides are targets for methylation at carbon atom 5. Only about 3% of the cytosines in human DNA are methylated, but most that are methylated are found in the CpG dinucleotide, producing 5-methylcytosine. Over evolutionarily long periods of time, 5-methylcytosine spontaneously deaminates to give thymine and so CpG is continuously being depleted and replaced by TpG (or CpA on the complementary strand. Despite the overall background, certain small regions of DNA noted for their transcriptional activity are characterized by the expected CpG density.

Base composition can also show regional subchromosomal variation. For example, human telomeres are defined by numerous repeats of a 50% GC sequence, TTAGGG. Large tracts of condensed heterochromatin found at centromeric regions and in defined noncentromeric regions of many chromosomes are composed of specialized repetitive DNA whose sequence can vary considerably in base composition from the bulk DNA. Such variation is the basis for the ability of equilibrium density ultracentrifugation to fractionate the total DNA into subclasses. The alternating pale and light staining chromosome bands also differ in a number of features, with dark G bands being characterized by a lower GC content than the pale staining bands.

Gene density in the human nuclear genome: The total number of genes in the human genome has been estimated to be about 70 000–80 000. As all but 37 of these genes are located in the nuclear genome, this gives a rough estimate of about 3000 genes per chromosome. However, gene density can vary substantially between chromosomal regions and also between whole chromosomes. For example, heterochromatic regions are known to be very largely composed of repetitive noncoding DNA, and the centromeres and large regions of the Y chromosome, in particular, are notably devoid of genes.

Recently, insight into gene distribution along the lengths of the different chromosomes has been obtained by hybridizing purified CpG island fractions of the genome (which are associated with perhaps about 56% of human genes; Antequara and Bird, 1993) to metaphase chromosomes (Craig and Bickmore, 1994). On this basis, it is clear that gene density is high in subtelomeric regions and that some chromosomes (e.g. 19 and 22) are gene rich while others (e.g. 4 and 18) are gene poor. Differences between pale and dark G bands, which are, respectively, gene rich and gene poor, are illustrated by the contrast between the human leukocyte antigen (HLA) complex and the dystrophin gene (*DMD*) regions. The former is located in the pale G band, 6p21.3: at the time of writing, the most intensively investigated region of the HLA complex, the class III region, had 70 genes in 0.9 Mb of DNA, giving a density of one per 13 kb. By contrast, a full 2.4 Mb of DNA in the dark G band region, Xp21, appears to be devoted exclusively to the dystrophin gene.

Organization and Distribution of Human Genes

The nuclear genome contains about 65 000–80 000 genes but only about 3% of the genome represents coding sequences

The number of genes in the human genome has been the subject of much speculation; while the small mitochondrial genome is known to have precisely 37 genes, the number in the nuclear genome remains unknown. Theoretical calculations based on the *mutational load* that a genome can tolerate and observed average mutation rates of human genes (~10^{-5} per gene per generation) suggest an upper limit of about 100 000. A variety of different approaches have been used to obtain more precise estimates of the total gene number. Three approaches have suggested a best estimate of about 65 000–80 000 genes:

- Genomic sequencing. Extrapolation from sequencing of large chromosomal regions may suggest that there are about 70 000 genes (Fields *et al.*, 1994). This is based on the observation that gene-rich regions have an average gene density of close to one per 20 kb, but gene-poor regions have a much lower density, say one-tenth of this density, and that the genome is split 50:50 into gene-rich and gene-poor regions.
- CpG island number. Restriction enzyme analysis using the methylation-sensitive enzyme *Hpa*II suggests that the total number of CpG islands in the human genome is 45 000 (Antequara and Bird, 1993). Using an estimate that approximately 56% of genes are associated with CpG islands, these authors have suggested a total of about 80 000 human genes.

- EST analysis. Large-scale random sequencing of cDNA clones provides so-called expressed sequence tags (ESTs, see Section 13.2.3). Comparison of known human EST sequences with a large set of different human genomic coding DNA sequences listed in sequence databases has suggested a figure of about 65 000 human genes (Fields *et al.*, 1994).

The above values suggest that the genes in the nuclear genome represent about 99.95% of the total number of cellular genes. If the average size of a human nuclear gene, including introns, is taken to be about 10–15 kb, this would mean that if the genes did not show overlaps, the total nuclear DNA occupied by genes would be about 70 000 × (10–15) kb or about 700–1050 Mb which corresponds roughly to about 25–35% of the genome. As the vast majority of nuclear genes encode polypeptides and the coding sequence required for an average size human polypeptide is taken to be about 500–600 codons, that is 1.5–1.8 kb, only about 3% of the nuclear genome (80–100 Mb of the 3300 Mb) would be expected to have a coding function.

RNA-encoding gene families often have numerous family members

While the great majority of human genes are expected to encode polypeptides, a significant minority encode mature RNA molecules of diverse function. The mitochondrial genome is exceptional in that 65% (24/37) of the genes encode RNA but even in the case of the nuclear genome about 5% of the genes, perhaps 3000–4000 genes in all, are expected to encode RNA molecules. In common with other cellular genomes a considerable variety of genes in the human genome are devoted to making mature RNA molecules which assist in the general process of gene expression. Some, notably rRNA and tRNA are involved in translation of mRNA.

In addition, many other RNA families are involved in reactions leading to maturation not only of mRNA but also of rRNA, tRNA and other RNA species, involving both cleavage reactions and basespecific modification reactions. In addition, several other RNAs have diverse functions.

Ribosomal RNA (rRNA) genes: There are multiple rRNA genes. In addition to the two mitochondrial rRNA molecules, the 28S, 18S and 5.8S cytoplasmic rRNAs are encoded by a single transcription unit which is tandemly repeated about 250 times, comprising five clusters of about 50 tandem repeats located on the short arms of human chromosomes 13, 14, 15, 21 and 22. In addition, the 5S cytoplasmic rRNA is encoded by several hundred gene copies in at least three clusters on the long arm of chromosome 1. The major rationale for the repetition of cytoplasmic rRNA genes is likely to be based on gene dosage: by having a comparatively large number of these genes, the cell can satisfy the huge demand for cytoplasmic ribosomes needed for protein synthesis.

Transfer RNA (tRNA) genes: These belong to a very large dispersed gene family, comprising more than 40 different subfamilies each with several members which encode the different species of cytoplasmic tRNA. In addition to multiple copies of genes specifying the individual cytoplasmic tRNA molecules, there are several defective gene copies (*pseudogenes*).

Small nuclear RNA (snRNA) genes: A heterogeneous collection of several hundred small nuclear RNA species are encoded by a large dispersed family of genes. Many of the snRNA species are uridine-rich and are named accordingly, e.g. U3 snRNA means the third uridine-rich small nuclear RNA to be classified. Individual species of RNA are associated with specific proteins to form ribonucleoprotein particles (RNPs). Some are known to be important in RNA splicing. A large subfamily of perhaps about 200 genes are present in the nucleolus, and have been termed small nucleolar RNA (snoRNA). They have important roles in specific cleavage reactions and base-specific modifications during maturation of ribosomal RNA (see Smith and Steitz, 1997).

Other RNA genes: Additional RNA genes encode functionally diverse products, including the 7SL RNA component of the signal recognition particle which is required for protein export and the RNA component of telomerase, the enzyme required to synthesize DNA at the telomeres. More recently, evidence has been obtained suggesting that certain RNA genes encode products that are important in gene regulation. An important example is the *XIST* gene. This gene is thought to be the major gene involved in initiating the process of X chromosome inactivation, being expressed exclusively from inactivated X chromosomes. No long open reading frames can be identified, and gene function is thought to be carried out through an RNA product by a mechanism that remains obscure.

In addition, several RNA genes have been found at a variety of chromosomal regions that are known to be imprinted (imprinted genes are normally expressed from a maternally inherited copy or a paternally inherited copy, but not both; see Section 8.5.4). For example, the *H19* gene contains five exons and is expressed to give a polyadenylated cytoplasmic RNA which does not however associate with ribosomes. It shows a restricted pattern of expression during early development (fetal and neonatal liver, visceral endoderm and fetal gut) and is imprinted, since only the maternally inherited allele is expressed. Its functional significance is, however, unclear.

Functionally similar genes are occasionally clustered in the human genome, but are more often dispersed over different chromosomes

As seen in the previous section, some families of RNA genes are clustered. In the case of polypeptide-encoding gene families, some genes encoding identical or functionally related products are clustered, but often they are dispersed on several chromosomes.

Functionally identical genes: A very few human polypeptides are known to be encoded by two or more identical gene copies. Often, these are encoded by recently duplicated genes in a gene cluster, as in the case of the duplicated α-globin genes. In addition, some genes on different chromosomes encode identical polypeptides. Examples include members of histone gene subfamilies.

Functionally similar genes: A large fraction of human genes are members of gene families where individual genes are closely related but not identical in sequence. In many such cases the genes are clustered and have arisen by tandem gene duplication, as in the case of the different members of each of the α-globin and β-globin gene clusters. Genes which encode clearly related products but which are located on different chromosomes are generally less related, as in the case of the α-globin and β-globin genes. However, in the case of the *HOX* homeobox gene family which consists of clusters of approximately 10 genes on each of four chromosomes, individual genes on different chromosomes may be more related to each other than they are to members of the same gene cluster.

In addition to the above, genes encoding closely related tissue-specific isoforms, or subcellular compartment-specific isozymes are often nonsyntenic (i.e. located on different chromosomes).

Functionally related genes: Some genes encode products which may not be so closely related in structure, but are clearly functionally related. The products may be subunits of the same protein or macromolecular structure, components of the same metabolic or developmental pathway, or may be required to specifically bind to each other as in the case of ligands and their relevant receptors. In almost all such cases, the genes are not clustered and are usually found on different chromosomes.

Human genes show enormous variation in size and internal organization

Size diversity: Genes in simple organisms such as bacteria are comparatively similar in size, and usually very short. By contrast, complex organisms such as mammals show wide variation in gene size, a feature found especially in human genes which can vary in length from hundreds of nucleotides to several megabases. The enormous size of some human genes means that transcription can be time-consuming. For example, the human dystrophin gene requires about 16 hours to be transcribed, and transcripts undergo splicing before transcription is completed (Tennyson *et al.*, 1995).

As one would expect, there is a direct correlation between the size of a gene and the size of its product, but there are some striking anomalies. For example, apolipoprotein B has 4563 amino acids and is encoded by a 45 kb gene while the dystrophin gene is 2.4 Mb in length and encodes a product in muscle cells of 3685 amino acids.

Diversity in internal organization: There is an inverse correlation between gene size and the proportion of the gene length which is expressed at the RNA level. A very small minority of human genes lack introns and are generally very small genes. For those that do possess introns, the exon content as a percentage of gene length tends to be very small in large genes. This does not arise because exons in large genes are smaller than those in small genes: the average exon size in human genes is about 200 bp and, although very large exons are known, exon size is comparatively independent of gene length. Instead, the explanation is due to the huge variation in intron lengths: large genes tend to have very large introns. The relationship between gene and intron length is not, however, without anomalies: the human type 7 collagen gene (*COL7A1*) is an intermediate size gene (31 kb) but has a total of 118 exons and an average intron size of only 188 bp (Christiano *et al.*, 1994). The extraordinary number of exons in *COL7A1* is not even matched by the number of exons in the giant 2.4 Mb dystrophin (*DMD*) gene.

Rare examples of overlapping genes and genes within genes are known in the human genome

Partially overlapping genes: The genes of simple organisms are generally more clustered than those in complex organisms. The average gene density in the human genome is about one per 40–45 kb of DNA. Assuming a mean size of, say, 10–15 kb, human genes should be separated by about 30 kb of nongenic DNA on average. By contrast, average gene densities in simple organisms are very much higher: roughly one per 1, 2 and 5 kb, respectively, for *E. coli*, *Saccharomyces cerevisiae* and *Caenorhabditis elegans*. Simple genomes such as those of certain phages and bacteria often show examples of partially overlapping genes which use different reading frames, sometimes from a common sense strand. The human mitochondrial genome is another example of a simple genome packed with genetic information and it too has an example of such overlapping genes.

Genes within genes: The small nucleolar RNA (snoRNA) genes are unusual in that the majority of them are located within other genes, often ones which encode a ribosome-associated protein or a nucleolar protein. Possibly this arrangement has been maintained to permit coordinate production of protein and RNA components of the ribosome (Tycowski *et al.*, 1993).

In addition to the snoRNA genes there are a few examples of other genes being located within the introns of larger genes, and in some cases the internal genes as well as the host genes are known to encode polypeptides. Three illustrative examples are:

- *The neurofibromatosis type I (NF1) gene.* Intron 27 of the *NF1* gene spans about 40 kb and contains three small genes, each with two exons which are transcribed from the opposite strand to that used for the *NF1* gene.
- *The factor VIII gene.* Intron 22 of the blood clotting factor VIII gene (*F8C*) contains a CpG island from which two internal genes, *F8A* and *F8B* are transcribed in opposite directions. *F8A* is transcribed from the opposite strand to that used by the factor VIII gene. *F8B* is transcribed in the same direction as the factor VIII gene to give a short mRNA containing a new exon spliced on to exons 23–26 of the factor VIII gene.
- *The retinoblastoma susceptibility gene RB1.* Intron 17 of this gene is 72 kb long and contains a G protein-coupled receptor gene, U16, which is actively transcribed from the opposite strand.

Human multigene families and repetitive coding DNA

DNA sequences in the nuclear diploid genome usually exist as two allelic copies (on paternal and maternal homologous chromosomes). In addition to this degree of repetition, about 40% of the human nuclear genome in both haploid and diploid cells is composed of sets of closely related nonallelic DNA sequences (DNA sequence families or repetitive DNA). Within the considerable variety of different repetitive DNA sequences are DNA sequence families whose individual members include functional genes (multigene families), and also many examples of nongenic repetitive DNA sequence families.

The reassociation kinetics of human DNA suggest three broad classes of DNA sequence

Reassociation kinetics first suggested that complex genomes, such as the human genome, comprise different sequence classes on the basis of the copy number. Typically, human DNA is randomly sheared (e.g. by sonication) to give fragments whose average size is about 500 bp and the sheared DNA is denatured by heating to separate the complementary strands of each fragment.

Thereafter the DNA is cooled to a temperature of about 20–30°C below the melting temperature, T_m (which marks the mid-point of the transition between the double-stranded and single-stranded states of DNA heated in solution). The cooled DNA renatures but the rate of reassociation depends not only on time (t) but also on the initial concentration (C_o) of that sequence (i.e. the C_ot value).

The above type of analysis has suggested that the human genome consists of three broad sequence components:

- *Single copy*, or at least very low copy number, DNA (60%) - reassociates very slowly. A single strand from a single copy sequence will require some considerable time to find a complementary partner strand, given that the vast majority of DNA fragments are unrelated to it.
- *Moderately repetitive* (30%) - intermediate speed of reassociation.
- *Highly repetitive* (10%) - reassociates very rapidly. There are numerous copies of the same sequence and the chances of quickly finding complementary partners within the mass of different fragments are high.

Members of DNA sequence families can be identified by a variety of different approaches.

The operational definition of a DNA sequence family is the comparatively high level of DNA sequence similarity (sequence homology) between whole family members, or components of the family members. Members of a DNA sequence family can be identified and actively sought by a variety of methods:

- *DNA sequencing* - Allows direct calculations on the degree of sequence relatedness of family members.
- *DNA hybridization and cloning.* A probe from a gene family member typically gives a complex band pattern when hybridized against a Southern blot of genomic DNAs. Individual family members can then be cloned by screening genomic DNA libraries.
- *PCR cloning* - Permits identification of novel family members by designing degenerate primers corresponding to highly conserved nucleotide or amino acid sequences.

When two members of a repetitive DNA sequence family exhibit a high degree of sequence homology, a recent common evolutionary origin is indicated. As detailed in the following sections, DNA sequence families show considerable variation in the number of different repeat unit members in the family, the size of the repeating unit, chromosomal location, mode of repetition and capacity for expression.

Human gene families vary in the overall sequence relatedness of different family members and the extent to which particularly conserved subgenic sequences define the family

A large percentage of actively expressed human genes are members of families of DNA sequences which show a high degree of sequence similarity. However, the extent of sequence sharing and the organization of family members can vary widely. Some family members may be nonfunctional (pseudogenes and gene fragments - see below) and rapidly accumulate sequence differences, leading to marked sequence divergence.

Classical gene families: Classical gene families are distinguished by members which exhibit a high degree of sequence homology over most of the gene length or, at least, the coding DNA component, a feature which automatically identifies such sequences as being closely related evolutionarily as well as functionally. In some cases, such as the individual rRNA gene families and individual histone gene families, there is an extremely high degree of sequence similarity between family members. Many other large gene families show a high degree of sequence similarity between family members.

Gene families encoding products with large, highly conserved domains: In some gene families there is particularly pronounced homology within specific strongly conserved regions of the genes; the corresponding sequence similarity between the remaining portion of the coding sequence in the different genes may be quite low. Often such families encode transcription factors that play important roles in early development, and the conserved sequence encodes a protein domain which is required to bind specifically to the DNA of selected target genes.

Gene families encoding products with very short conserved amino acid motifs: The members of some gene families may not be very obviously related at the DNA sequence level, but nevertheless encode gene products that are characterized by a common general function and the presence of very short conserved sequence motifs:

- the DEAD box gene family - contains several different genes whose products appear to function as RNA helicases and are characterized by the presence of eight short amino acid sequence motifs, including the DEAD box (the sequence Asp-Glu-Ala-Asp as represented by the one-letter amino acid code);
- the WD repeat gene family - gene products have a regulatory function, but there is considerable diversity in function. Typically characterized by tandem repeats with a central core of fixed length and containing small conserved amino acid motifs, including the WD (tryptophan-aspartate) sequence);
- the ankyrin repeat gene family - wide functional diversity but often involved in protein-protein interactions. Characterised by tandem repeats of a 33 amino acid sequence characterized by the presence of select amino acids at only a few positions;

- the LIM domain gene family - encode a characteristic cysteine-rich 56 amino acid domain most likely involved in protein-protein interactions.

Gene superfamilies: In some types of gene family, the genes encode products that are known to be functionally related in a general sense, and show only very weak sequence homology over a large segment, without very significant conserved amino acid motifs. Instead, there may be some evidence for general common structural features. Such genes, which appear to be evolutionarily related but more distantly than those in a classical or conserved domain/motif gene family, have been considered to be members of a gene superfamily.

For example, in addition to the immunoglobulin gene family, other related genes such as the *HLA* genes, TCR genes, *T4* and *T8* genes are known to encode products with an immune system function and a domain structure that resembles that of immunoglobulins. Although, therefore, the level of sequence homology between such genes may be very low, the similarities in function and general domain structure have suggested the existence of a so-called Ig superfamily, in which there appears to be a distant common evolutionary relationship.

Human gene families can occur as closely clustered genes at specific subchromosomal locations, or as widely dispersed genes

A wide variety of human gene families have been identified and show considerable variation in both the organization of the genes and the extent to which individual genes within the gene family are related in terms of sequence and function. In terms of gene organization, two basic arrangements can be discerned: families where there is evidence of close gene clustering and families that are dispersed over several different chromosomal locations. This classification is rather arbitrary, however, since several gene families consist of multiple gene clusters at different chromosomal locations.

Gene families organized in a single cluster: Genes in an individual gene cluster are thought to arise by tandem gene duplication events. Different organizations are evident:

- Tandem gene organization. This arrangement is exemplified by the organisation of genes in individual rRNA gene clusters. The genes are highly related to each other in terms of both sequence and function, although certain family members may be nonfunctional.
- Close clustering. The individual genes may be more physically separate than in a tandemly repeated cluster but nevertheless may be closely clustered, even to the extent of being subject to a common regulatory mechanism (locus control region - see the example of the α- and β-globin gene clusters. Again the individual genes usually show a high degree of sequence and functional identity to each other, but many family members may be pseudogenes.
- Compound clusters. In other clustered gene families, however, the physical relationship between genes in a cluster may be less close and a cluster of related genes may also contain within it genes that are unrelated in sequence and function, a compound gene cluster. For example, the HLA complex on 6p21.3 is dominated by families of genes which encode class I and II HLA antigens and various serum complement factors, but individual family members may be separated by functionally unrelated genes such as members of the steroid 21-hydroxylase gene family, etc. The gene organization in the latter case shows evidence of an ancestral duplication event that involved a segment containing different genes that are unrelated in function and sequence.

Gene families organized in multiple gene clusters

Some gene families are organized in clusters distributed over two or more chromosomal locations. Again different organizations are evident, with some families showing very high similarity between genes on different clusters, while others are marked by comparatively low sequence homology between genes on different clusters.

- High cluster similarity. Here sequence homology between gene family members on different chromosomes may be very high as in the case of the different rRNA gene clusters on the short arms of chromosomes 13, 14, 15, 21 and 22. Another useful example is the olfactory receptor gene family which encodes a diverse repertoire of receptors which allow us to discriminate thousands of different odors. The family consists of perhaps 1000 genes, making it the largest in the human genome, and is organized in large clusters at more than 25 different chromosomal locations.
- Low cluster similarity. Often sequence homology is greater within a cluster than between clusters. For example, the globin gene family includes genes at three locations: the α-globin cluster on 16p, the β-globin cluster on 11p and the myoglobin gene at 22q and although all globin genes are clearly related, those within a cluster are more related to each other than they are to genes in one of the other clusters.

Interspersed gene families: Some gene families show no obvious physical relationship between family members, which are dispersed as solitary genes at two or more different chromosomal locations. The family members may show considerable sequence divergence unless their dispersion has been a relatively recent event, or there has been considerable selection pressure to maintain sequence conservation. The following examples illustrate some of the many different types of organization.

- *Families expected to originate from two genomes*. Current ideas on mitochondrial origins have envisaged that the original genome was derived from a prokaryote but that many of the genes were subsequently transferred to the nuclear genome. Some families of nuclear genes encode cytoplasm-specific and mitochondrial-specific isoforms for certain enzymes and other key metabolic products (see *Table 7.5* for some examples).
- *Families originating from ancient genome duplication or gene duplication events*. This class of interspersed gene families

typically contains only a few members, as in the case of the *PAX* gene family, and appear to have evolved by a combination of gene duplication and/or genome duplication events over a long period of evolutionary time. Typically, all or most of the genes are functional, and may individually encode highly related products.

- *Families originating largely by retrotransposition events.* Some gene families have expanded comparatively recently in evolutionary terms by a process whereby RNA transcribed from one or a small number of functional genes is converted by cellular reverse transcriptase into natural cDNA which then becomes integrated elsewhere in the chromosomes. Most such copies are nonfunctional.

Pseudogenes, truncated gene copies and gene fragments are commonly found in multigene families

Families of RNA genes or polypeptide-encoding genes are frequently characterized by defective copies of essentially all of the gene or its coding sequence (pseudogenes), or a portion of it, in some cases a single exon (gene fragments). A large variety of different classes are found. The following examples are meant to be illustrative of the types of defective gene copies found in different types of gene family.

Nonprocessed pseudogenes in a gene cluster: Individual gene clusters are typically characterized by the presence of defective gene copies which have been copied *at the level of genomic DNA*. This means that the defective gene copies can contain copies of the exons, introns and promoter regions of the functional genes (nonprocessed pseudogenes). Classical examples are the α-globin and β-globin clusters.

Only one member of the α-globin gene family is a nonprocessed pseudogene, but three of the seven gene copies in the β-globin gene cluster are. Another β-globin gene family member, *HBQ1,* is likely to be an expressed pseudogene: it encodes a type of globin polypeptide, θ-globin, but there is no evidence that the latter is ever incorporated into a haemoglobin molecule and so θ-globin is likely to lack any function.

Truncated genes and gene fragments in a gene cluster: The class I HLA gene family at 6p21.3 is a classical example of a gene cluster which is characterized by nonprocessed pseudogenes, truncated gene copies and gene fragments. Although the number of class I *HLA* genes can vary on different chromosome 6s, comprehensive analysis of one of these identified 17 family members clustered over about 2 Mb. Six of the genes are known to be expressed, although the precise functions of some of these are still not clearly understood.

The remaining members are clearly defective. Four are conventional full-length pseudogenes, but another five represent truncated gene copies (lacking the 5' end in four cases and the 3' end in the other case) and two are fragments which contain a small component of the gene sequence, even a single exon.

Nonprocessed pseudogenes in an interspersed gene family: For several polypeptide-encoding genes, defective copies containing intronic sequence have been found elsewhere in the genome. Two illustrative examples are:

- *The* NF1 *(neurofibromatosis type I) gene.* This gene is located close to the chromosome 17 centromere at 17q11.2 and has at least 11 nonprocessed pseudogene or gene fragment copies, nine of which are located at pericentromeric regions on seven different chromosomes (Regnier *et al.*, 1997). Characterization of the chromosome 15-specific *NF1* gene copies revealed that they contain copies of a segment of the *NF1* gene spanning exons 8 and 27 and including the intron sequences. This type of gene duplication may be a result of what has been called pericentromeric plasticity.
- *The* PKD1 *(adult polycystic kidney disease) gene.* This gene consists of 46 exons spanning 50 kb at 16p13.3. A truncated 5' gene copy comprising approximately 70% of the gene and encompassing exons 1 to 34 and all the intervening introns has been faithfully replicated at least three times and inserted into a more proximal location at 16p13.1 (the European Polycystic Kidney Disease Consortium, 1994).

An apparently related case concerns the *CFTR* (cystic fibrosis) gene at 7q31.2. which has 24 exons spanning 250 kb. Copies of a ~30 kb sequence encompassing exon 9 and flanking intron sequences are distributed at a large number of chromosomal locations. However, in this case a likely explanation proposed by Rozmahel *et al.* 1997 is a retrotransposition model. Transcription from the sense strand of the *CFTR* gene was envisaged to generate an antisense transcript encompassing the exon 9 sequence and flanking intron regions. The transcript could then have been converted to cDNA by cellular reverse transcriptase prior to integration at several sites in the genome.

Processed pseudogenes in an interspersed polypeptide-encoding gene family: Interspersed gene families frequently have several defective gene copies which have been copied *at the cDNA level.* This form of retrotransposition is carried out by cellular reverse transcriptases which transcribe mRNA into natural cDNA which can then integrate into chromosomal DNA at sites of temporary breakage. Although processed pseudogenes are typically not expressed, several examples are known where natural cDNA copies of a gene appear to be expressed, often in a testis-specific fashion. Often the functional intron-containing gene locus in such families is located on the X chromosome. *Processed pseudogenes in an RNA-encoding gene family:* Although the size of some interspersed polypeptide-encoding gene families testifies to the success of retrotransposition as a mechanism for generating processed gene copies, the really successful (in terms of high copy number) retrotranspositions have been performed from RNA polymerase III transcripts. For example, the Alu repeat family is considered to have arisen as processed pseudogenes copied from the 7SL RNA gene. Genes such as this which are transcribed by RNA polymerase III often contain an internal promoter which facilitates the expression of newly transposed copies.

The coding sequences of many human genes contain repeated sequence motifs

In addition to the types of gene sequence duplication discussed in the previous sections, many human genes, like other eukaryotic genes, contain intragenic repeated sequences. Repeated sequences in coding DNA may involve different forms of repeated structure, including comparatively large sequences encoding protein domains or small sequence motifs. Different modes of repetition can be seen but tandem repetition is rather common.

- Tandem repetition of microsatellite sequences is common and may simply reflect statistical expected frequencies for certain base compositions. For example, certain genes which have a very high % GC often have long runs of single cytosines or guanines in the coding sequence. Such sequences are comparatively unstable and are prone to single nucleotide deletion or insertion events which are often pathogenic. A variety of genes are known to have long tracts of certain trinucleotides in their coding sequences and runs of the CAG trinucleotides in the coding sequences of several genes have been found to be capable of expansion causing pathogenesis.
- Tandem repetition of sequences encoding known or assumed protein domains is quite common, and may be functionally advantageous in some cases by providing a more available biological target. In some cases, the sequence homology between the repeats can be very high; in other cases it may be rather low.

Extragenic Repeated DNA Sequences and Transposable Elements

The human nuclear genome, like that of other complex eukaryotes, contains a large amount of highly repeated DNA sequence families which are largely transcriptionally inactive. A wide variety of different repeats are known and dedicated repeat sequence databases have been established. Like multigene families, noncoding repetitive DNA shows two major types of organization: tandemly repeated and interspersed.

Tandemly repeated noncoding DNA: Such families are defined by blocks (or arrays) of tandemly repeated DNA sequences. Individual arrays can occur at a few or many different chromosomal locations. Depending on the average size of the arrays of repeat units, highly repetitive noncoding DNA belonging to this class can be grouped into three subclasses: satellite, minisatellite and microsatellite DNA.

In addition to these three major subclasses (which are detailed in the following three sections), a fourth class has recently been recognized and described as megasatellite or macrosatellite DNA. Despite the name, this type of DNA is characterized by array lengths which can be comparatively modest compared to some satellite DNA arrays. Instead, the prefix mega- has been used to emphasize the large size of the repeating unit which can be several kilobases long. This class is exemplified by the RS447 megasatellite which consists of about 60 tandem copies of a novel 4.7 kb repeat on 4p15, plus another array of several copies on distal 8p. Array lengths can be highly polymorphic and the RS447 repeat is reported to contain a putative open reading frame of 1590 bp.

Interspersed repetitive noncoding DNA: The individual repeat units are not clustered, but are dispersed at numerous locations in the genome, and together account for perhaps one third of the DNA in the human genome (Smit, 1996). Most of the DNA families belonging to this class contain some members that are capable of undergoing retrotransposition (i.e. transposition through an RNA intermediate).

The chromosomal locations of different types of tandemly repeated DNA can show a very restricted or highly dispersed pattern, whereas different classes of interspersed repeat DNA can show preferential location within different types of chromosome bands.

Satellite DNA is composed of very long arrays of tandem repeats which can be separated from bulk DNA by buoyant density gradient centrifugation

Human satellite DNA is comprised of very large arrays of tandemly repeated DNA with the repeat unit being a simple or moderately complex sequence. Repeated DNA of this type is not transcribed and accounts for the bulk of the heterochromatic regions of the genome, being notably found in the vicinity of the centromeres (pericentromeric heterochromatin). The base composition, and therefore density, of such DNA regions is dictated by the base composition of their constituent short repeat units and may diverge substantially from the overall base composition of bulk cellular DNA.

Isolation by buoyant density gradient centrifugation: When DNA is isolated from human cells by conventional methods, it is subject to mechanical shearing. Fragments are generated from the bulk DNA (with a base composition of ~42% GC) and fragments from the satellite DNA regions which may have a similar or different base composition. If the base composition is significantly different, satellite DNA sequences can be separated from the bulk DNA by buoyant density gradient centrifugation. Following centrifugation, they appear as minor (or satellite) bands of different buoyant density from a major band which represents bulk DNA. Typically, human DNA is complexed with Ag^+ ions and then fractionated in buoyant density gradients containing cesium sulfate, whereupon three satellite bands are identified at different densities: satellite 1 – 1.687 gcm^{-3}; satellite 2 – 1.693 gcm^{-3}; satellite 3 – 1.697 gcm^{-3}.

Each of these satellite classes includes a number of different tandemly repeated DNA sequence families (satellite subfamilies), some of which are shared between different classes. DNA sequence analysis has revealed that some of the repetitive DNA families in the satellites are based on very simple repeat units. For example, both satellite 2 and satellite 3 contain sequence arrays which are based on tandem repetition of the sequence ATTCC. Additionally, restriction mapping has revealed satellite subfamilies which show additional higher order repeat units superimposed on the small basic repeat units. Such subfamilies are thought to arise as a result of subsequent amplification of a unit which is larger than the initial basic repeat unit and contains some diverged units.

Alphoid DNA and centromeric heterochromatin: Other types of satellite DNA sequence cannot easily be resolved by density gradient centrifugation. They were first identified by digestion of genomic DNA with a restriction endonuclease which typically has a single recognition site in the basic repeat unit. In addition to the basic repeat unit size (monomer), such enzymes will produce a characteristic pattern of multimers of the unit length because of occasional random loss of the restriction site in some of the repeats (Singer, 1982a). Alpha satellite (or alphoid DNA) constitutes the bulk of the centromeric heterochromatin and accounts for about 3–5% of the DNA of each chromosome. It is characterized by tandem repeats of a basic mean length of 171 bp, although higher order units are also seen. The sequence divergence between individual members of the alphoid DNA family can be so high that it is possible to isolate chromosome-specific subfamilies for each of the human chromosomes (Choo *et al.*, 1991).

The function of satellite DNA remains unclear. The centromeric DNA of human chromosomes largely consists of various families of satellite DNA. Of these, only the α-satellite is known to be present on all chromosomes, and its repeat units often contain a binding site for a specific centromere protein, CENP-B. Recently cloned α-satellite arrays have been shown to seed *de novo* centromeres in human cells, indicating that α-satellite plays an important role in centromere function.

Minisatellite DNA is composed of moderately sized arrays of tandem repeats and is often located at or close to telomeres

Minisatellite DNA comprises a collection of moderately sized arrays of tandemly repeated DNA sequences which are dispersed over considerable portions of the nuclear genome. Like satellite DNA sequences, they are not normally transcribed (but see below).

Hypervariable minisatellite DNA: Hypervariable minisatellite DNA sequences are highly polymorphic and are organized in over 1000 arrays (0.1–20 kb long) of short tandem repeats (Jeffreys, 1987). The repeat units in different hypervariable arrays vary considerably in size, but share a common core sequence, GGGCAGGAXG (where X = any nucleotide), which is similar in size and in G content to the *chi* sequence, a signal for generalized recombination in *E. coli.* While many of the arrays are found near the telomeres, several hypervariable minisatellite DNA sequences occur at other chromosomal locations.

The great majority of hypervariable minisatellite DNA sequences are not transcribed, except for elements occuring within noncoding intragenic sequences. Some, however, are expressed. For example, the *MUC1* locus on 1q is an expressed hypervariable minisatellite locus. It encodes a glycoprotein found in several epithelial tissues and body fluids which is highly polymorphic as a result of extensive variation in the number of minisatellite-encoded repeats (Swallow *et al.*, 1987).

The significance of hypervariable minisatellite DNA is not clear, although it has been reported to be a 'hotspot' for homologous recombination in human cells (Wahls *et al.*, 1990). Nevertheless it has found many applications. Various individual loci have been characterized and used as genetic markers, although the preferential localization in subtelomeric regions has limited their use for genome-wide linkage studies. A major application has been in DNA fingerprinting, in which a single DNA probe which contains the common core sequence can hybridize simultaneously to multiple minisatellite DNA loci on all chromosomes, resulting in a complex individual-specific hybridization pattern.

Telomeric DNA: Another major family of minisatellite DNA sequences is found at the termini of chromosomes, the telomeres. The principal constituent of telomeric DNA is 10–15 kb of tandem hexanucleotide repeat units, especially TTAGGG, which are added by a specialized enzyme, telomerase. By acting as buffers to protect the ends of the chromosomes from degradation and loss and by providing a mechanism for replicating the ends of the linear DNA of chromosomes, these simple repeats are directly responsible for telomere function.

Microsatellite DNA is defined by the presence of short arrays of tandem simple repeat units and is dispersed throughout the human genome

Microsatellite DNA families include small arrays of tandem repeats which are simple in sequence (often 1–4 bp) and are interspersed throughout the genome. Of the mononucleotide repeats, runs of A and of T are very common and together account for about 10 Mb, or 0.3% of the nuclear genome. By contrast, runs of G and of C are very much rarer. In the case of dinucleotide repeats, arrays of CA repeats (TG repeats on the complementary strand) are very common, accounting for 0.5% of the genome, and are often highly polymorphic. CT/AG repeats are also common, occurring on average once every 50 kb and accounting for 0.2% of the genome, but CG/GC repeats are very rare.

This is so because C residues which are flanked at their 3' end by a G residue (i.e. CpG) are prone to methylation and subsequent deamination, resulting in TpG (or CpA on the opposite strand). Trinucleotide and tetranucleotide tandem repeats are comparatively rare, but are often highly polymorphic and increasingly have been investigated to develop highly polymorphic markers.

The significance of microsatellite DNA is not known. Alternating purine-pyrimidine repeats, such as tandem repeats of the dinucleotide pair CA/TG, are capable of adopting an altered DNA conformation, Z-DNA, *in vitro*, but there is little evidence that they do so in the cell. Although microsatellite DNA has generally been identified in intergenic DNA or within the introns of genes, a few examples have been recorded within the coding sequences of genes. Tandem repeats of three nucleotides in coding DNA may be sites that are prone to pathogenic expansions.

Highly repeated interspersed DNA families contain a small percentage of actively transposing DNA elements

Two major classes of mammalian interspersed repetitive DNA families have been discerned on the basis of repeat unit length (Singer, 1982b): SINEs and LINEs.

SINEs (short interspersed nuclear elements): The most conspicuous human SINE is the *Alu* repeat family (so called because of early attempts at characterizing the sequence using the restriction nuclease *Alu*I). The *Alu* repeat contains an internal RNA polymerase III promoter sequence. It has attained a very high copy number in the human genome and appears to have originated by retrotransposition from the 7SL RNA gene. The *Alu* repeat is primate-specific but other mammals have similar types of sequence derived from the 7SL RNA gene such as the B1 family in mouse.

Unlike the *Alu* repeat, another major human SINE family is not restricted to primates, with copies being found in marsupials and monotremes. In accordance with its distribution this family has been termed the MIR (mammalian-wide interspersed repeat) family. *LINEs (long interspersed nuclear elements):* Human LINEs are exemplified by the *LINE-1* or *L1* element (also called the *Kpn* repeat because of early attempts at characterizing this family using the restriction nuclease *Kpn*I). The LINE-1 element is also found in other mammals such as the mouse. In addition to the human *Alu* and LINE-1 repeat families, there are many smaller families, including the THE-1 (transposable human element family), many MER (medium reiteration frequency) families and families of human endogenous retroviruses (HERV) or retrovirus-like elements (RTLV).

Some members of the interspersed repeat families have been considered as transposable elements, unstable DNA elements which can migrate to different regions of the genome. Rare examples are known of human DNA sequences which appear to have been copied by a DNA-mediated transposition event (transposons) and a variety of sequences have been found in the human genome which resemble known DNA transposons such as the *mariner* transposon and others (Smit and Riggs, 1996). However, the latter are not thought to be actively transposing. Instead, the great majority of human transposable elements undergo retrotransposition, that is their RNA transcripts can be converted within the cell to a complementary DNA form which can reinsert back into chromosomal DNA at a variety of different locations. Three classes of mammalian sequence are known to be able to transpose through an RNA intermediate. In each case, the transposition event involves duplication of a very short sequence at the target site, causing the transposed sequence to be flanked by short repeats.

The Alu repeat occurs about once every 3 kb in the human genome and includes examples that are transcribed

The *Alu* repeat is the most abundant sequence in the human genome, with a copy number of about 1 000 000 (see Deininger,

1989; Smit, 1996). *Alu* repeats have a relatively high GC content and, although dispersed mainly throughout the euchromatic regions of the genome, have been reported to be preferentially located in R chromosome bands (Korenberg and Rykowski, 1988). The latter correspond to the pale bands seen when using standard Giemsa staining and represent the most transcriptionally active regions of the genome.

The full-length *Alu* repeat is about 280 bp long and is usually flanked by short (often 6–18 bp) *direct repeats* (i.e. the repeats are in the same orientation). The typical *Alu* sequence is a tandemly repeated dimer, with the repeats sharing an approximately 120 bp sequence followed by a short sequence which is rich in A residues on one strand and T residues on the complementary strand. However, there is asymmetry between the tandem repeats: one repeat unit contains an internal 3'-bp sequence lacking in the other. Monomers, containing only one of the two tandem repeats, and various truncated versions of dimers and monomers are also common.

The two repeated units of the *Alu* sequence show a striking resemblance to the sequence for 7SL RNA, a component of the signal recognition particle, which facilitates transport of proteins across the membrane of the endoplasmic reticulum.

Because of this and the observation of the A-rich regions and the flanking direct repeats, it has been widely assumed that the *Alu* sequence has been propagated by retrotransposition from 7SL RNA, and therefore represents a processed 7SL RNA pseudogene. Certainly, transposition by *Alu* sequences is known to occur [presumably as a result of *trans*-acting cellular reverse transcriptases such as those encoded by LINE-1 (*Kpn*) elements, and may occasionally cause clinical problems]. Possibly, the very high copy number achieved by this processed pseudogene is related to the presence of a promoter sequence in the 7SL RNA sequence (the 7SL RNA gene, like tRNA genes is transcribed by RNA polymerase III from an internal promoter). By contrast, processed pseudogenes from RNA polymerase II transcripts lack promoter sequences and their only chance of expression is if the integration event places them next to a functional promoter sequence.

Currently, the function of the *Alu* sequence, if any, is unknown, although several roles have been considered (see Schmid, 1998). Although the average expected frequency is one copy per 3 kb, high density clustering of *Alu* repeats is known to occur in certain regions. Because of their ubiquity, *Alu* sequences have been considered to promote unequal recombination, a mechanism which while occasionally causing disease, may be evolutionarily advantageous in promoting gene duplication.

Although conspicuously absent from coding sequences, *Alu* sequences are often found in noncoding intragenic locations, notably in introns and occasionally in untranslated sequences. Consequently, they are often represented in the primary transcript RNA from genes encoding polypeptides, and occasionally in mRNA. The *Alu* sequence can also be transcribed from its internal promoter by RNA polymerase III *in vitro*, and *in vivo* transcription of some *Alu* sequences can result in the accumulation of a small cytoplasmic RNA which can be specifically bound by two cytoplasmic signal recognition particle proteins, SRP9 and SRP14 (Chang *et al.*, 1994).

The LINE-1 (Kpn repeat) includes examples that appear to encode a reverse transcriptase and are actively transposing

The human LINE-1 (L1, *Kpn* repeat) family is expected to consist of over 100 000 and perhaps as many as 400 000– 500 000 interspersed repeats (Smit, 1996). Of these, multiple members are known to be actively transposing (Sassaman *et al.*, 1997).

The full-length consensus element is 6.1 kb long and has two open reading frames (ORFs). ORF1 is located close to one end (conventionally known as the 5' end) of the consensus element and encodes a protein of unknown function, p40 (so called because its molecular weight is ~40 kDa). ORF2 encodes a protein with an endonuclease domain and also a reverse transcriptase domain (see Sassaman *et al.*, 1997 ; Kazazian and Moran, 1998). The full-length consensus element contains an internal promoter within a region of untranslated DNA preceding ORF1 (conventionally called the 5' -UTR) while at the other end there is an $(A)_n/(T)_n$ sequence, often described as the 3' poly(A) tail. As in the case of other elements that transpose, the LINE-1 elements are flanked by short duplicated repeats.

The full-length LINE-1 element is comparatively rare and most repeats are truncated at the 5' end, resulting in a population that is heterogeneous in length but sharing a common 3' end with the poly(A) tail. LINE-1 elements are primarily located in euchromatic regions but show an inverse relationship with *Alu* repeats by appearing to be located preferentially in the dark G bands (Giemsa positive) of metaphase chromosomes. Like the *Alu* repeats, they are conspicuously absent from coding sequences but may be found in intragenic noncoding sequences. As a result, they may be represented in the primary RNA transcript of large genes, but they are conspicuously absent from coding sequences.

BIBLIOGRAPHY

Adelberg, Edward A.: *Papers on Bacterial Genetics*, Boston, Brown and Company, 1960.

Ausubel, F.M.: *Current Protocols in Molecular Biology*, New York, John Wiley and Sons, 1989.

Bailey, Ronald: *Liberation Biology: The Scientific and Moral Case for the Biotech Revolution*, Amherst, Prometheus Books, 2005.

Barnes, N.: *Biology*, New York, Worth Publishers, 1989.

Barnum, Susan R.: *Biotechnology: An Introduction*, Belmont, Thomson/Brooks/Cole, 2005.

Bauer MW: *Biotechnology-the Making of a Global Controversy*, Cambridge, Cambridge University Press, 2002.

Bernauer, Thomas: *Genes, Trade, and Regulation: The Seeds of Conflict in Food Biotechnology*, Princeton, Princeton University. 2003.

Brandwein, P.F.: *Sourcebook for the Biological Sciences*, San Diego, Harcourt Brace JOvanovich, 1986.

Broach, J.R.: *The Molecular Biology of the Yeast*, Cold Spring Harbor, Cold Spring Harbor Laboratory, 1981.

Brooks, Rodney A.: *Flesh and Machines: How Robots Will Change Us*, New York, Pantheon Books, 2002.

Chrispeels, Maarten : *Plants, Genes and Crop Biotechnology*, Sudbury MA, Jones and Barlett Publishers, 2003.

Clark, J.M.: *Experimental Biochemistry*, New York, W.H. Freeman and Company, 1977.

Cox, Earl D. : *Beyond Humanity: Cyber Evolution and Future Minds*, Roackland, Charles River Media, 1996.

Curtis, H.: *Biology*, New York, Worth Publishers, 1989.

David Sadava: *Plants, Genes and Crop Biotechnology*, Sudbury MA, Jones and Barlett Publishers, 2003.

DeGregori, Thomas R.: *Bountiful Harvest: Technology, Food Safety and the Environment*, Washington DC, Cato Institute, 2003.

Dixon, Dougal: *After Man-A Zoology of the Future*, New York, St. Martin's Press, 1981.

Drexler, K. Eric: *Engines of Creation: The Coming Era of Nanotechnology*, New York, Anchor Books, 1986.

Fransman M, Junne G, Roobeek A: *The Biotechnology Revolution?*, Oxford, Blackwell, 1995.

Freidberg, Susanne: *French Beans and Food Scares: Culture and Commerce in an Anxious Age*, New York, Oxford University Press. 2004.

Friedberg, E.C.: *DNA Repair*, New York, WH Freeman and Company, 1985.

Fukuyama F.: *Our Posthuman Future: Consequences of the Biotechnology Revolution*, NewYork, Farrar, Straus and Giroux, 2002.

Fumento, Michael: *Bioevolution: How Biotechnology is Changing Our World*, San Francisco, Encounter Books, 2003.

Gaskell G: *Biotechnology-the Making of a Global Controversy*, Cambridge, Cambridge University Press, 2002.

George S. Paul: *Beyond Humanity: Cyber Evolution and Future Minds*, Roackland, Charles River Media, 1996.

Goodman, D.C.: *From Farming to Biotechnology: A Theory of Agro-industrial Development*, Oxford, Blackwell, 1987.

Goodsell, David S.: *Bionanotechnology: Lessons From Nature*, Hoboken, Wiley-Liss, 2004.

King, R.B., Mihony, G., and Sheldon, J.K.: *Practical Environmental Bioremediation*, Lewis Publishers, Boca Raton, 1992.

Kurzweil, Ray: *The Age of Spiritual Machines*, New York, Penguin Books, 1999.

Miller, Henry I. : *The Frankenfood Myth: How Protest and Politics Threaten the Biotech Revolution*, New York, Praeger, 2004.

Miller, R.V.: *Strategies and Mechanisms for Field Research in Environmental Bioremediation*, American Academy of Microbiology, Washington, D.C., 1994.

Moravec, Hans: *Mind Children: The Future of Robct and Human Intelligence*, Cambridge, Harvard University Press, 1988.

Morholt, E. : *Sourcebook for the Biological Sciences*, San Diego, Harcourt Brace JOvanovich, 1986.

Murray, David: *Seeds of Concern: The Genetic Manipulation of Plants*, Sydney, University of New South Wales, 2003.

Old, R.W. : *Principles of Gene Manipulation*, London, Blackwell Scientific Publications, 1989.

Parens, Erik: *Enhancing Human Traits: Ethical and Social Implications*, Washington, D.C.: Georgetown University Press, 1998.

Poindexter, J.S.: *Strategies and Mechanisms for Field Research in Environmental Bioremediation*, American Academy of Microbiology, Washington, D.C., 1994.

Postman, Neil: *Technopoly: The Surrender of Culture to Technology*, New York, Vintage Books, 1992.

Primrose, S.B.: *Principles of Gene Manipulation*, London, Blackwell Scientific Publications, 1989.

Retzer, W.J.: *Biotechnology Workbook*, Englewood Cliffs, Prentice Hall, 1991.

Rifkin, Jeremy: *The Biotech Century*, New York, Penguin Putnam, 1998

Sayler, G.S., Fox, R., and Blackburn, J.W.: *Environmental Biotechnology for Waste Treatment*, Plenum Press, New York, 1991.

Shetty, Kalidas: *Food Biotechnology*, New York, Dekker/CRC Press, 2005.

Slater, R.J.: *Experiments in Molecular Biology*, Clifton, Humana Press, 1986.

Smith, John E.: *Biotechnology*, Cambridge, Cambridge University Press, 2004.

Switzer, R.L.: *Experimental Biochemistry*, New York, W.H. Freeman and Company, 1977.

Towle, Albert: *Modern Biology*, Austin, Holt, Rinehart and Winston, 1988.

Walden, Richard: *Genetic Transformation in Plants*, England, Open University Press, 1988.

INDEX

O

P

R

S

T

V

□□□